ECLIPSE

The celestial phenomenon that changed
the course of history

DUNCAN STEEL

Foreword by Paul Davies

The Joseph Henry Press
Washington, DC

Joseph Henry Press • 2101 Constitution Avenue, N.W. • Washington, D.C. 20418

The Joseph Henry Press, an imprint of the National Academy Press, was created with the goal of making books on science, technology, and health more widely available to professionals and the public. Joseph Henry was one of the founders of the National Academy of Sciences and a leader in early American science.

Library of Congress Cataloging-in-Publication Data

Steel, Duncan, 1955–
  Eclipse : the celestial phenomenon that changed the course of history / Duncan Steel.
      p. cm.
Includes bibliographical references and index.
  ISBN 0-309-07438-X (alk. paper) *2726 6005 6/03*
  1. Eclipses—Popular works. I. Title.
QB541 .S65 2001
523.7′8—dc21

                              2001039904

Cover: August 1999 solar eclipse as seen from Neunkirchen, Austria, © Reuters/Heinz-Peter Bader/Archive Photos; eclipse sequence photo © 1999 by Fred Espenak, courtesy of www.MrEclipse.com.

For my parents, Ken and Shirley

Contents

Foreword

By now it was about midday and there came a darkness over the whole land, which lasted until three in the afternoon; the sun was in eclipse. And the curtain of the temple was torn in two.

This quotation from the gospel of Luke (quoted from the New English Bible) describes one of the most momentous celestial events in history, for it reportedly occurred at the time of Jesus Christ's death. Whether or not this is an accurate account, or a literary embellishment, it well illustrates the deep significance that all ancient cultures have attached to solar eclipses.

When the Moon passed across the face of the Sun on August 11, 1999, the event was watched avidly by tens of millions of people all over Europe and the Middle East. Some saw it as a millennial sign, others expected miracles, while the majority simply enjoyed this natural spectacle. It happens that eclipses repeat on a regular cycle, with gaps of 18 years plus 10 or 11 days. The next one in that particular sequence, occurring on August 21, 2017, will sweep its path of totality across more than a dozen of the United States, the first total solar eclipse to visit North America for 38 years (although Hawaii got one in 1991). If you can't wait that long, there are several opportunities elsewhere–such as in southern Africa and Australia at the end of 2002–but you would need to travel in order to experience the unique feeling such an event evinces.

Total solar eclipses always generate huge interest from the general public. Partly this reflects a deep-seated mystical appeal that attaches to these astronomical phenomena, partly it is because of their sheer aesthetic beauty. The glaring solar disk that is so familiar to us swamps the weak light from the tenuous gas that envelops the Sun. Only during an eclipse is the delicate tracery of this material revealed, in the form of the solar corona. For a few precious minutes the sky is suffused with the eerie coronal glow, a sight that must have struck awe and terror into the hearts of our ancestors.

In astronomical terms, however, an eclipse is a perfectly straightforward affair. By a strange coincidence, the angular sizes of the Sun and Moon as viewed from Earth are almost exactly the same. The Sun is, of course, a much bigger body, but located a lot farther away. If the Moon's orbit around the Earth were in the same plane as the Earth's orbit around the Sun, eclipses would occur every month, but because the orbital planes are tilted obliquely relative to one another, only rarely do the Earth, Moon, and Sun stand in alignment. Even when that happens, the Moon's shadow projected onto the Earth's surface is so small that, at any given spot, total eclipses can be hundreds of years apart. For example, before 1999 the last one to occur anywhere in mainland Britain was in 1927; preceding that, one has to go back to 1715 and 1724. Now there's not another one due to visit those islands until 2090. This rarity factor adds considerably to the sense of excitement.

There is no simple formula to predict the dates of eclipses, and it took heroic effort on behalf of priests and astrologers to figure out when they were likely to occur. It is a testimony to the enormous supernatural significance bestowed upon eclipses by early cultures that the complex mathematical rules were worked out independently so many times. From the Mayas to the Sumerians, from the Egyptians to the Chinese, algorithms were devised to foretell the next time the Sun would be gobbled up and the sky would go dark in daytime.

These early attempts to predict eclipses amounted to no more than fitting cycle times to observations using trial and error. No physical understanding lay behind them. By the seventeenth century, however, Isaac Newton had formulated his laws of motion and gravitation, and the prediction of eclipses became a classic application of Newtonian physics. At last human beings could comprehend why solar and lunar eclipses happened when they did. Today, with computer models of planetary motion, the dates for future eclipses can readily be worked out for millennia ahead. Conversely, astronomers can "retrodict," and find out when eclipses occurred in history. In turn, this can be checked against ancient records, allowing history and astronomy to confirm each other.

Historically, solar eclipses have provided ideal opportunities for scientists to study a range of phenomena besides the spectacular corona. For example, by timing the exact moment of transit, the Moon's position can be measured to very high precision. Although the Moon orbits the Earth in a roughly elliptical path, there are small variations, such as a very gradual drift away from Earth. It is important for astronomers and geophysicists to understand these corrections. Today, better measurements can be made using laser ranging.

One of the most famous scientific uses of a solar eclipse occurred on May 29, 1919. Four years before, Albert Einstein had published his general theory of relativity, which explains gravitation in terms of a warping or curvature of space-time. Among the various manifestations of these geometrical distortions is a minute "spacewarp" around the Sun. In effect, the Sun's gravitational field acts like a lens, slightly bending light beams that pass close to the solar surface. Einstein computed that a star beam might be deflected in this way by up to 1.75 seconds of arc, thus displacing the apparent position of the star in the sky when it is located close to the Sun along our line of sight. This is actually a very tiny shift, though measurable with a good telescope. Unfortunately, the Sun's glare prevents us from observing stars in the daytime, but during a total solar eclipse, the stars become visible. The British astronomer Sir Arthur Eddington went on an expedition to the island of Principe off Spanish Guinea in Africa, with the express purpose of testing Einstein's prediction. The results brilliantly vindicated the general theory of relativity, and more than anything else served to propel Einstein to international fame. For some decades, this "bending of light" measurement constituted one of only three firm tests for the theory, which is regarded by many as the intellectual triumph of the twentieth century.

Today solar eclipses have somewhat less scientific value than in the past, but they still attract an enormous amount of attention, not just from keen astronomers, but also from the public at large. Some people are so inspired by observing a total eclipse that they will travel thousands of miles to see another. Although this unique astronomical phenomenon is a consequence of the normal cycles and rhythms of the Solar System, and need no longer be feared as a bizarre supernatural sign, the splendor and spectacle of the event ensures that the popularity of solar eclipses will never wane.

Paul Davies

Preface

Total solar eclipses are so infrequent one might say they happen "once in a blue moon." That phrase is used colloquially to imply something seldom occurring, but here the allusion is a mixed metaphor: eclipses, of course, involve the Moon, and it looks pretty black when obscuring the Sun during such an event.

How often does a blue moon occur? That's an impossible question to answer, because there are several distinct—even contradictory—meanings for the phrase. This is something we'll discuss in detail later, but for the present let us assume the modern definition of two full moons within a single calendar month. Full moons occur about 29 and a half days apart, so you could get two within a 30-day month, but this is much more likely in a 31-day month. It is quite straightforward to show that a blue moon, according to that definition, happens about once every 32 or 33 months, on average.

Now, how does that compare with the frequency of eclipses? In fact there are at least two eclipses every year, and there may be up to seven, but that includes lunar as well as solar eclipses, the majority being partial events (incomplete shadowing). Total solar eclipses occur *somewhere* on Earth on average once per 18 months, but very often they happen over the oceans or the poles. Half of the eclipse tracks might cross some populated land, so that the frequency of potential visibility is typically once per 36 months (unless you are keen enough to take a trip to the Antarctic to catch one).

Even that is less than the blue moon frequency, but more pertinent is how often a total solar eclipse track traverses a particular location on our planet. How long do you need to wait until one of these visits your city and state? The answer is that a random location on the Earth is graced by such an event once every four centuries, once every five or six human lifetimes. Blue moons happen all the time, compared to that.

Blue moons can tell us something else about eclipses and their frequency. The year 1999 was unusual for several astronomical reasons, such as an eclipse over Europe (discussed below), a transit of Mercury across the face of the Sun, and the recurrence of the great Leonid meteor shower. It was also a double blue moon year. Because February had only 28 days, it happened to have no full moon in 1999, whereas the adjacent January and March had two each.

Now step forward a few decades. Calculations show that in both 2018 and 2037 we may anticipate double blue moons, again in January and March. One immediately notices that there are 19-year gaps. Full moons are spaced by lunar months. Counting up those months, there are 235 in each interval.

Looking backwards, 1961 was a double blue moon year. With this information in hand you might bet that 1980 was also a double blue moon year, but it was not: only March contained two full moons. A clue as to why that was the case comes from the fact that 1980 is divisible by four: it was a leap year. Our sequence of 19-year gaps between double blue moon occurrences was upset by the way we choose to correct for the length of the solar year not being an exact number of days long. That is, a period of 19 *solar* years is close to 235 lunar months long, but any particular 19 *calendar* years will vary by a day because there can be either four or five leap years contained therein. In 1980 a full moon slipped out of January into the lengthened February.

This period of 19 solar years or 235 lunar months is called the Metonic cycle. The Christian churches use it to calculate the dates of Easter; Judaic clerics employ it to define the Hebrew calendar; and many other cultures frame their annual rounds with the Metonic cycle as their basis. The cycle gives us a handle on when blue moons may occur, a mere curiosity. More significantly, it allows eclipses to be predicted.

The total solar eclipse in 1999, on August 11, was eagerly awaited. It was the first to cross any part of Britain for 72 years. After touching down in the Atlantic near Newfoundland, the track of totality proceeded eastwards to the southwestern tip of England where I (along with many others) was waiting for it. The Moon's shadow then swept onwards across France, Germany, several eastern European countries, Turkey, the Middle East, Pakistan, and India, before eventually petering out in the Bay of Bengal. Millions and millions of people experienced its effects, many having traveled around the globe knowing full well what was to happen. Looking up the tables, there were also eclipses on August 11, 1961 and August 10, 1980 (the leap year upset the date again), and another is due on August 11, 2018. Obviously the Metonic cycle produces some eclipse regularity. In this book we will see that there are also several other systematic features of eclipses, allowing their prediction by knowledgeable people. Nowadays that information is easy to find, but step back a few centuries or millennia, when eclipses were viewed variously as augurs of ill or harbingers of good fortune: the ability to prophesy eclipses would surely have brought great power and influence. Eclipse dates are clearly inter-twined with the calendar—to a surprising extent, we will discover.

Herein I describe not only solar and lunar eclipses (and their celestial brethren such as transits and occultations by planets, com-ets, and asteroids), but also the great influence these events have had upon the advance of civilization. Knowing when eclipses were due enabled more scientific societies to gain an advantage over others, a matter discussed in the opening chapter. To appreciate how these cosmic events could be predicted by the ancients, long before Nicolaus Copernicus described how the planets orbit the Sun, or Sir Isaac Newton expounded his law of gravity, one needs to understand the cycles and systematics of eclipses, matters dis-cussed in detail in the Appendix. Some may find this heavy going (although it involves only simple arithmetic), and that is why it appears at the end. If you really want to comprehend the astro-nomical cycles involved, read the Appendix first. For most readers, though, the information in the opening chapters will be sufficient. After that we delve into the cultural and scientific importance of eclipses, in the distant and nearer past, and the future.

Now some words about my sources of information. Many of the eclipse computations used, plus the map shown as Figure 2-5, are derived from the excellent Internet site of Fred Espenak, who works at NASA-Goddard Space Flight Center in Greenbelt, Maryland. Anyone who wants to know more is strongly recom-mended to take a look at Espenak's pages, making a start at: http://sunearth.gsfc.nasa.gov/eclipse/eclipse.html

In the age of the Internet, I have accessed several hundred web sites in preparing this book. Especially because so many of these are ephemeral, there is no point in listing them. Similarly I have made use of some dozens of books to obtain information, to greater and lesser extents. Any interested reader will easily find a multitude of popular-level books and magazine articles dealing with various aspects of eclipses. A few are listed among the picture credits near the end of this volume. A core subject here is the cause of eclipses and their cycles, as described in the Appendix but with related considerations being scattered throughout the text. This is not a matter often treated in books suitable for the non-specialist. Therefore I mention here that my description is based largely on the detailed analysis that appears in the book *Eclipses of the Sun and Moon* by Frank Dyson and Richard Wooley, published by the Clarendon Press, Oxford, England, in 1937.

Many people have kindly answered questions for me, or helped me with photographs and other illustrations. In particular I would like to mention Graeme Waddington, Tony Beresford, John Hisco, Fraser Farrell, David Asher, Bill Napier, Brian Marsden, Daniel McCarthy, Peter Davison, John Kennewell, Alain Maury, Philippe Veron, Leslie Morrison, Steven Bell, Jim Klimchuk, and Paul Davies.

ECLIPSE

1

From the Depths of Time:
The Earliest Recorded Eclipses

Zeus, the father of the Olympic Gods, turned mid-day into night, hiding the light of the dazzling Sun; and sore fear came upon men.

Archilochus, referring to the total solar eclipse of 648 B.C.

Eclipses have had a profound and startling effect upon the cultural development of humankind. Let us begin by asking which Eclipse has exerted the greatest influence over our affairs.

In this opening chapter we will describe various famous historical eclipses, such as those that presaged several great battles in antiquity, interpreted by one side as an auspicious omen, by the other as a portent of doom. We will also mention the eclipse that seems to have followed the death of Jesus Christ on the cross, and left a strong impression upon His followers and foes alike. But in my opening paragraph I was not asking about any such eclipse in the sky.

With a little sleight of hand, I capitalized the word *Eclipse* there. The most famous Eclipse of all time was an eighteenth-century British racehorse by that name, which happened to be born at the time of a solar eclipse visible in England in 1764. He continues to affect everyday life because every thoroughbred carries a few of his genes.

Never beaten in a race, after retirement from the track Eclipse spent almost 20 years at stud. After his death in 1789, the great horse's skeleton was mounted at the Royal Veterinary College in London, at least one hoof was turned into a snuffbox, and in several countries there are annual races called the Eclipse Stakes. In the United States a set of annual awards recognize the top thoroughbreds in various categories, and these are called the Eclipse Awards. For the actual horse, then, it is a postmortem mixture of abasement and honor.

But those are trivialities. Is horseracing as a whole so important as to justify my claim? While I am not an aficionado of the so-called Sport of Kings, I recognize its significance. In terms of economic turnover, horseracing and the associated activities (like gambling) are reckoned to represent one of the largest industries in many nations. In Britain, Ireland, and France it is an especially large slice of the economy. One has only to visit the Kentucky Derby or the Happy Valley racecourse in Hong Kong to see that the equine Eclipse has a continuing sway upon human activity, 200 years after his death.

If you look up the word "eclipse" in a dictionary of quotations, among the entries the following will often appear: "Eclipse first, and the rest nowhere." Those words are often uttered as a prognosis on any sporting contest in which the outcome is a foregone conclusion. Dennis O'Kelly, the owner of Eclipse, famously coined the phrase when he wagered that he could place the first three horses in a race.

CELESTIAL SHOWTIME

When it comes to celestial displays, "Eclipse first, and the rest nowhere" also represents a succinct summary of the opinion of the eclipse enthusiast. Many people routinely book flights and accommodation a year ahead to ensure they are in the right place at the right time to experience a total solar eclipse, then feverishly check the next scheduled performance in this free-to-view continuing astronomical extravaganza to begin planning their next trip.

Is it really a case of "the rest nowhere" when it comes to heavenly displays? I think not, and would argue that a meteor storm (when the sky briefly lights up with myriad shooting stars for perhaps an hour) is not only more spectacular, but also less frequent, occurring only once per decade or so. But there is something that distinguishes a total solar eclipse.

All those on the correct side of the planet may witness a meteor storm, when a stream of tiny comet-derived rocks intercepts the Earth. Every year there are several distinct meteor showers, half a dozen of which are worth watching for even the casual observer, such as the Perseids on August 12 and the Geminids around December 13. Every so often a greater concentration of meteors is anticipated, as with the Leonid meteor storm (keep watching the sky in the early hours of November 17/18 for each of the next few years). But the shooting stars may be seen from a large area of the planet, whereas the track of a total solar eclipse is narrow, leading to exclusivity. Would diamonds be regarded so highly if they were common, to be found under every upturned boulder?

A total solar eclipse occurs through the combination of several unlikely circumstances. It happens that the angular diameters of the Sun and the Moon as viewed from the Earth are about the same. Those apparent sizes vary, though, because the distance from the Earth to the Sun changes during the year, and the separation between the Moon and us also oscillates each month. To get totality, the Moon must be near enough and the Sun far enough such that the lunar disk can completely block the Sun. The next condition is that the Moon must cross the plane of the Earth's orbit very close to the direction of the Sun. If that happens then the Moon's shadow is cast somewhere onto the Earth and a partial solar eclipse occurs over a wide area, but complete blocking—a total eclipse—is experienced only within a narrow band by a fortunate few. A total solar eclipse occurs somewhere around the globe about once every 18 months, but as the track of totality is usually less than 100 miles wide, you should expect to wait for several centuries in any random location for the next one. Such eclipses have rarity value, a prize worth chasing around the globe, and many people do just that, chalking up their eclipses in exotic locations and showing anyone interested their best photographs of the event.

There is another type of eclipse: a lunar eclipse. These occur when the terrestrial shadow envelops the Moon, and again it may be either total or partial. Unlike solar eclipses, the viewing constraints are not so stringent: you just need to be somewhere on the night side of the Earth at the appropriate time, and half the human race might see a particular lunar eclipse.

Solar eclipses have been interpreted as evil omens by many civilizations because the life-giving sunlight is obscured for a few minutes, producing a profound effect upon all under the celestial shadow. Lunar eclipses, although they last far longer, are not so unmistakable. Despite this fact, such eclipses have also been taken very seriously indeed by many societies because rather than going black, the Moon instead darkens to the color of blood as it is eclipsed, instilling fear and dread into the superstitious witness.

Lunar eclipses are not prized for their scarcity, but because they are seen often they provided the opportunity for early civilizations to build an understanding of the basic cycles, leading to detailed predictions of when and where eclipses (both solar and lunar) were due.

THE APPEARANCE OF ECLIPSES

Hackneyed though the thought may be, it is true that a picture is worth a thousand words. Let's stop our verbal description of eclipses and take a look at some pictures. We need to familiarize ourselves with the beasts that are the lead characters (I can hardly call them "stars") in our story.

The Sun is not simply a bright yellow disk in the sky. Actually, it is a hugely complicated heat-generating engine that solar physicists are still a long way from understanding completely. Car bumper stickers say "Solar Power Not Nuclear Power," but solar power *is* nuclear power. The Sun is a gigantic nuclear reactor, fusing hydrogen nuclei together to produce helium and liberating vast amounts of energy in consequence.

Nor is the Sun constant. Its appearance alters substantially, with sunspots being seen much of the time, moving across its face as it rotates. The numbers and forms of sunspots vary over an activity cycle lasting a little more than 11 years. Figure 1-1 shows two extreme examples of the Sun's guises, with no major sunspots at the minimum of its cycle, but an extremely pockmarked profile

FIGURE 1-1. Two views of the Sun that show how it can change. Near the minimum of the solar activity cycle (left), no major blemishes are seen. Close to the maximum (right) the Sun appears pockmarked with sunspots.

being presented at the maximum. At such a time—the last peak was in 2000—the Sun tends to jettison larger amounts of matter through the outward-flowing stream of charged particles called the solar wind, resulting in vivid auroras being seen on Earth and sometimes in the disruption of radio communications.

Similarly the Moon is not quite so simple as one might think. Figure 1-2 shows two full moons. They differ in several respects. Firstly, the apparent size of the lunar disk alters because its orbit about our planet is noncircular, and when the Moon is closer to us it appears bigger. One of these full moons happened to be when it was near perigee, making the angular size of the disk larger; the other was at apogee, with the result that it appears smaller.

Secondly, the brightness distributions over the two images are not quite the same. This has various causes. One is that full moon

FIGURE 1-2. Contrasting full moons seen near perigee (left) and apogee (right) indicate how much the apparent size of the Moon varies each month.

occurs when the Sun, Earth, and Moon are aligned in terms of the celestial longitude (looking down from above they are in a straight line), and say 12 hours before that point the Moon may *appear* full, but the alignment is not yet precise. Another reason for sunlight reaching the lunar surface at an angle, producing brightness variations, is that full moon generally occurs either above or below the Earth's orbital plane. (If the Moon crosses that plane at full moon then a lunar eclipse results.)

Thirdly, various wobbles cause more than just 50 percent of the lunar surface to be visible from Earth. Notice that the two vistas shown in Figure 1-2 are somewhat different.

Since we are concerned here with eclipses, next some eclipse images. Figure 1-3 shows a total solar eclipse, the Moon obscuring the whole solar disk. Some sunlight can still be seen through

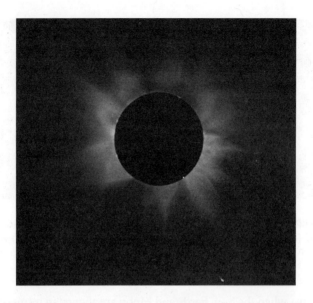

FIGURE 1-3. A total solar eclipse photographed in perfect conditions. The large bright areas are coronal streamers. Close to the solar disk, bright prominences are visible, in particular on the left; these appear pink in real life.

various mechanisms. Around the circumference of the Moon is the *corona*, the term derived from the Latin and thence Spanish word meaning "crown" or "garland." This is a solar structure, far beyond the Moon, made visible only during an eclipse. Basically it is the tenuous glowing atmosphere of the Sun. The corona appears white to the eye. Vividly seen here are various coronal streamers.

Another distinct phenomenon that may be glimpsed during a total solar eclipse is the existence of transitory *prominences*, huge red clouds of glowing gas thrown up above the solar surface. These are also apparent in Figure 1-3, as bright spots close to the edge of the Moon. These prominences are often more obvious to the naked eye than in a photograph, and old-time eclipse-watchers were very familiar with them, terming them the "red flames." An example of this is the nineteenth-century sketch of totality shown in Figure 1-4. Before the advent of photography, this was the only way to record what was seen.

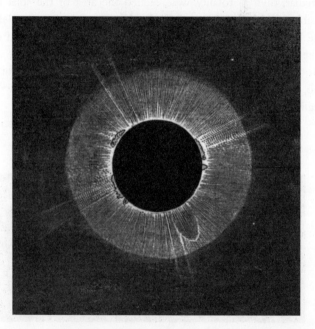

FIGURE 1-4. A total solar eclipse sketched by a nineteenth-century astronomer. To the naked eye the prominences jutting above the solar surface often appear more noticeable than they are in a photograph.

Figure 1-5 shows an eruptive prominence on a grand scale, without the benefit of an eclipse. This blob of superheated gas, 80,000 miles long, was thrown off the Sun in 1996 and its behavior was captured in ultraviolet light with a sensor on board the SOHO (Solar and Heliospheric Observatory) satellite. This series of images stretches over five hours, the prominence moving outwards at more than 15,000 miles per hour.

If the profiles of both the Moon and the Sun were perfectly circular, then as totality was reached the area of the solar disk visible would diminish until finally there was just one spot left in the line of sight, and then that would be abruptly blanked out. The effect would be that a dim corona would surround the dark Moon, with a single bright spot momentarily witnessed before it, too, disappeared. This so-called *diamond ring effect* is shown in Figure 1-6.

In reality neither the Sun nor the Moon is precisely circular,

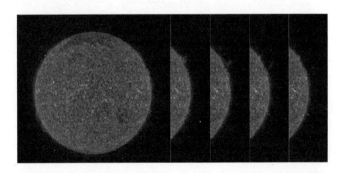

FIGURE 1-5. A five-hour sequence of ultraviolet images of the Sun obtained with the SOHO satellite show a developing eruptive prominence at upper right.

FIGURE 1-6. The diamond ring effect is seen just as totality starts and ends. This shows an example, but it's a bit of a cheat. This was an artificial eclipse of the Sun by the Earth, photographed by the Apollo 12 astronauts on their way back from the Moon late in 1969, produced when their spacecraft slipped into the terrestrial shadow.

and under some circumstances a series of bright spots may also be seen around the limb. These are due to light squeezing through the valleys between craters and mountains at the edge of the Moon. These spots are called *Baily's beads* after the nineteenth-century British astronomer Francis Baily, who first described their form and origin in 1836. Baily's sketch is shown in Figure 1-7. The edge of the Moon is certainly crinkled, as one can see in Figure 1-8; during an eclipse the mountains of the Moon are cast in sharp relief, as seen in Figure 1-9.

We have seen how the basic aspects of solar eclipses appear in the sky; let us now return to our discussion of how humans have reacted to such events over the eons.

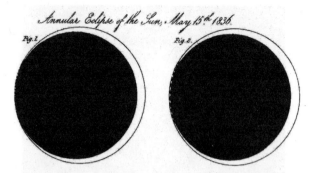

Annular Eclipse of the Sun, May 15th 1836.

Fig. 1.   Fig. 2.

FIGURE 1-7. Baily's beads, from his original description in 1836.

FIGURE 1-8. The lunar limb is not a smooth arc, but is lined with mountains and craters as shown in this photograph obtained by the Apollo 11 astronauts in orbit around the Moon in 1969.

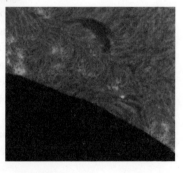

FIGURE 1-9. The mountainous edge of the Moon is obvious in this segment photographed during an eclipse. Note also the appearance of the solar surface.

EARLY HUMAN EXPERIENCES OF ECLIPSES

How long have members of our species witnessed eclipses and wondered at their origin and implications? This is not a question we can answer with assurance, but we can guess.

Sequences of marks scratched on animal bones dating back 30,000 years are suggestive of the changing phases of the Moon from one cycle to the next. Certainly the varying brightness of that orb as it waxes and wanes would be important if you were reliant upon it to find your way at night, as was all humankind for most of our history. The snuffing out of moonlight for several hours when it was expected to be full would be a matter of some concern to such people. That's what happens during a total lunar eclipse. The fact that the Moon turns the color of blood would also leave a strong impression, making observers speculate about the significance of such an episode.

Lunar eclipses occur at a rate of about 15 per decade, a little less than half being total (that is, the entire lunar disk is enveloped by the Earth's shadow). All inhabitants of the side of the planet facing the Moon would be able to witness it, as long as clouds do not intervene.

Even if a typical human back then lived for only 30 years, still some dozens of lunar eclipses would have been seen by each individual, and primeval societies must have been familiar with them. Remember that early humans did not live in cities with artificial lighting, so they were much more attuned to the sky, and dependent upon its cycles. Nowadays rural people are rather more aware of celestial events than city-dwellers, and in the past the average person's acquaintance with the firmament above was developed to a greater extent than it is today.

To ancient peoples a lunar eclipse would provoke some consternation, but a solar eclipse would be even more amazing, and worrying. Total solar eclipses intrinsically occur more often, but may be seen from only a restricted part of the globe because the track of the lunar shadow drawn across the ground is typically only 60 to 100 miles wide. Outside that track, the solar eclipse is partial, dimming the sunlight but not obstructing it altogether. In a typical decade there are seven or eight total solar eclipses, but you have to be in the right place at the right time to experience any of them. The zone of partiality may cover half of one side of our planet, but the startling totality track is much more restricted.

ANCIENT ECLIPSE RECORDS

The records of early civilizations are littered with references to eclipses. This should not be a surprise, in that only the most notable events in any year, or decade, or even century, will have been remembered by later generations and recorded for posterity in some way. The sorts of things passed down to us would be human signposts such as the births and deaths of kings, and the great battles and wars in which they were involved, plus unusual natural phenomena like disastrous floods and earthquakes and the appearances of bright comets and eclipses. The mass media of today are swamping us with the trivialities of life; the documents of the past, painstakingly chiseled into rock or inked onto papyrus scrolls, were limited to only the most prominent events.

One of the great early civilizations was that of the Babylonians, and their knowledge of astronomy is especially notable (see Figure 1-10). The ebbs and flows of that people, and the changes that occurred in the lands around the mighty Tigris and Euphrates

FIGURE 1-10. Ancient Babylonian astronomers discuss the appearance of a comet, as the Moon rises in the east. They developed a comprehensive understanding of eclipse cycles.

rivers, greatly affected humankind's eventual understanding of eclipses. The Babylonians scratched their records into clay tablets and these have proven a unique repository of information for studies of ancient eclipses.

It was not only the Babylonians who were interested in eclipses. Elsewhere other civilizations were in awe of such events. In China, India, Arabia, ancient Greece, and medieval Europe, eclipses were seen and not forgotten; their dates and characteristics were written down and stored, invaluable records to be translated much later and put to disparate scholarly uses.

CALIBRATING CALENDARS USING ECLIPSES

When we know how to convert the dates of ancient eclipses to our modern calendar, the records provide useful information about the Earth's spin history, and the trend over the past 2,700 years is now well delineated. But the converse is also true. If one has a definite report that an eclipse was recorded from a certain place in a certain year then one can calibrate the calendar the locals were using.

Consider the calendar of the Roman Republic. After about 400 B.C. the Romans were using a scheme whereby in most years there were 12 months adding up to 355 days, rather than the actual solar year which averages close to 365.25 days. That leaves a deficit of over ten days a year. Every so often the leaders of the Senate were supposed to declare an additional month of 22 or 23 days, to be inserted into February, but they were quite lax in this regard for various reasons. One was that the thirteenth month was considered unlucky, hence the common fear of the number 13 (triskaidekaphobia). The major reason, though, was that they were able to manipulate the year length to their own advantage for taxation or electoral purposes. The outcome was that the date on the Roman calendar seldom bore any clear relation to the season. The end of May would come, but still it was winter. Looking back it would be almost impossible for historians to be able to say what occurred when, if it were not for eclipses.

In 168 B.C. the Romans defeated the Greeks in the Battle of Pydna, a town on the western side of the Gulf of Salonika. The battle was pivotal for the eventual Roman control of Greece because it quelled the Macedonians. (These were the people who in the latter half of the fourth century B.C. had produced Alexander the Great, and under him conquered an empire stretching from the eastern end of the Mediterranean all the way to India.) The Romans recorded this battle as occurring, on their haphazard calendar, on September 3. From this, one might imagine that it took place in early fall. In fact we know that the Battle of Pydna was fought near midsummer's day. The great writers Livy and Pliny recorded that a lunar eclipse was predicted by the tribune Sulpicius Gallus and seen on the night before the conflict, giving courage to one side while the other was filled with dread. Perhaps the Roman generals chose the date on the basis of the prediction, telling their troops ahead of time that there would be an eclipse as a sign of divine favor.

Knowing that the year was 168 B.C. we are able to back-calculate the date of the eclipse and check its visibility from Greece. Using the regular calendar introduced by Julius Caesar over a century later, and projecting it backwards, the eclipse was on June 21. This nicely confirms the basis of the story related by Livy and Pliny: on a sensible, well-regulated calendar the battle was indeed fought close to midsummer's day.

From that eclipse we find that the Roman republican calendar in that year was 74 days out of synchronization with the seasons. The discrepancy at times had been even greater. A solar eclipse in 190 B.C. shows that the calendar then was ahead of the seasons by 119 days. When Julius Caesar introduced his eponymous calendar to begin in 45 B.C. (our modern calendar is the same except with a slight adjustment to the frequency of leap-year days), he had to add 80 days to 46 B.C. to make up for the shortcomings of his predecessors.

THE ADVANTAGE OF ECLIPSE FOREKNOWLEDGE

It seems incongruous that the Romans profited from the eclipse of 168 B.C., because their opponents, the Greeks, were much more proficient in scientific matters. The Greek knowledge of eclipses was largely derived not internally, but from the Babylonians after 330 B.C. The Babylonians had been subsumed into the empire built by Alexander the Great when he defeated the Persians, who had been occupying Mesopotamia through much of the fourth century B.C. as part of their own empire.

After conquering Egypt, Alexander had marched east and pushed the Persians out of Babylonia, pursuing them north into Assyria. When he eventually caught up with King Darius III and defeated him in a decisive battle at Gaugamela in 331 B.C., it was on the day after a lunar eclipse. Alexander interpreted this as an omen blessing the Greek endeavor. One wonders how he knew the eclipse was due. One possibility is information provided by the expert Babylonian astronomers. Perhaps they found Alexander and his people a preferable occupier to the Persians.

ECLIPSES IN THE BIBLE

Total solar eclipse tracks are narrow. To be able to say with surety, looking far back in time, that a certain eclipse was visible from a specific location requires that we know how the spin of the planet has varied over the past several millennia. Although we only have definite eclipse records dating back to 700 B.C., the trend can be extrapolated for perhaps another thousand years. This opens the possibility of identifying dates for a few of the eclipses mentioned in the Old Testament.

One of the best-known allusions to an eclipse occurs in the book of Genesis. Referring to Abraham in Canaan, the text says, "And when the Sun was going down . . . great darkness fell upon him." It is possible to identify this description with a computed solar eclipse occurring on May 9, 1533 B.C., which would have occurred at about 6:30 P.M. local time (indeed, when the Sun was going down).

In the biblical account, a great comet was seen the following year, a fact in itself of interest to modern astronomers. To investigate the dynamical history of comets, long temporal baselines of observations are important. For Halley's Comet, observations back to 240 B.C. are known, and earlier records would be useful. The comet of 1532 B.C. is not linked with any recently observed object, but maybe it will reappear soon.

The most famous solar eclipse in the Bible is that of Joshua. This has long puzzled scholars, because it describes the Sun as stopping still during an eclipse, and even moving backwards. In terms of a date, this appears to have been the solar eclipse of September 30, 1131 B.C. But in regard to the phenomenon reported (the Sun halting or retreating), we are pretty sure that

Joshua was wrong. Similar claims have been made for several other eclipses, for example one that affected a fifteenth-century civil war in Ireland. This is merely a visual illusion produced by the Moon overtaking the Sun, the latter seeming to slip backwards in consequence.

Various events around 760 B.C., culminating in a major earthquake that damaged Solomon's Temple in Jerusalem and caused a huge destructive wave in the Sea of Galilee, clearly had a considerable effect upon the people of Judea. The fact that the rumble was felt over 800 miles away allows us to estimate that the tremble was about magnitude 7.3 on the Richter scale. In the Bible there are eight separate allusions to a solar eclipse around that time, which may be identified as June 15, 763 B.C. We are told the eclipse was followed closely by a bright comet. If this was Halley's Comet (we cannot be sure due to the sparsity of the information) then we know from a backwards extrapolation of its orbit that the comet indeed would have been visible in August of 763 B.C., five centuries before the earliest definite observation cited above. The combination of these pieces of biblical information leaves us pretty sure that the earthquake happened four years later, in 759 B.C.

That year's identification stems from our ability to back-calculate the eclipse. Clearly eclipses (and periodic comets) are important phenomena in that the modern understandings of astronomers and mathematicians enable historians to assign definite dates to events in the distant past. For example, when was Jesus crucified?

THE CRUCIFIXION ECLIPSE

Astronomers have long speculated about how their science might fix the date of the birth of Jesus, given the Star of Bethlehem story.

My own favored dating involves multiple conjunctions of the planets Jupiter, Saturn, and Mars that we know occurred in 7 and 6 B.C. The Magi would therefore have been alerted to some impending event—the coming of the Savior had long been awaited by the Jews—which suspicion was confirmed in their minds by the appearance of a bright comet early in 5 B.C. We know from Chinese records that the comet was visible for over 70 days and moved across the sky in accord with the Magi following it from the environs of Babylon first westwards to Jerusalem, and then the final handful of miles south to Bethlehem. Various other aspects of the biblical story fit in with this picture and lead to a deduction of the Nativity occurring in mid-April of 5 B.C.

But what of the *end* of the mortal life of Jesus, when he was crucified in his thirties? When did that melancholy event take place? Various commentators have discussed how the range of possible dates can be restricted, based upon facets mentioned in the Gospels, such as the temporal relationship of the Crucifixion to Passover. Because Passover is at full moon, and the Crucifixion was on a Friday, only certain dates are feasible. The chief candidates are April 7, A.D. 30, and April 3, A.D. 33.

The essential clue of an eclipse was missed until quite recently. In 1983 Colin Humphreys (now at Cambridge University) and Graeme Waddington (of Oxford University) recognized that the date of the Crucifixion might be identified in this way. They noted that in various places in the Bible, and other early written accounts, allusions are made to the Moon being dark and "turned to blood" when it rose in the evening after the Crucifixion, which sounds like a lunar eclipse. Mention is also made of the *Sun* being darkened earlier that day. This may have been due to a dust storm caused by the *khamsin*, a hot wind from the south that blows through the region for about 50 days commencing around the middle of March, in accord with the expected time of year. Such dust storms and their sun-dimming effects are well known.

Under such circumstances—a lunar eclipse whilst there was much dust suspended in the air—one would expect the Moon to appear the dark crimson of blood. With that in mind Humphreys and Waddington computed the dates of all lunar eclipses possibly visible from Jerusalem between A.D. 26 and 36. And they found one on April 3, A.D. 33, one of the two possible dates mentioned above, occurring as the Moon rose a couple of hours after Jesus died, in accord with the Gospels.

This all relates back to the quote from the New English Bible with which Paul Davies began his Foreword to this book. Over the eons, memories of events in quite separate years can become confused or melded in the mind and then are written down in a way that deviates from historical reality. The Gospels were not written until the last decades of the first century A.D., a generation and more after the period described, and since then successive copying and translation have moved away from the original text.

It happens that the Sun was *not* in eclipse at the time of the Crucifixion, but would have been darkened by the khamsin dust. This is obvious simply from knowledge of Jewish custom: Passover is at full moon, and the definite biblical link between the Crucifixion and Passover makes a lunar eclipse the only possibility. On the other hand, there *had* been a solar eclipse thereabouts in recent times. On November 24, A.D. 29, the path of totality had passed just north of Jerusalem, over Damascus, Beirut, and Tripoli. Jerusalem itself would have been severely darkened. This would have left a strong local memory, but the date excludes it from being at the time of the Crucifixion: the year is too early, and Passover is not in November. It seems clear that over some decades the memory of lunar eclipse and dust-dimmed Sun in A.D. 33 became combined with the total solar eclipse in A.D. 29, leading to the false idea that the Sun was eclipsed at the Crucifixion. The computed lunar eclipse, then, allows us to allot a date to the Crucifixion. It was on the third day of April in the year A.D. 33.

ECLIPSES AS PORTENTS OF DOOM

The ancients often interpreted eclipses as omens. To a modern-day rationalist, the notion that an eclipse could be a celestial portent for things to come would seem absurd. Nevertheless they *could* have an effect, through psychological action: men buoyed by belief in their righteousness bolstered by an eclipse are more likely to win in combat against those who take the eclipse to augur evil. Self-fulfilling prophecies do exist.

For example, a lunar eclipse in 413 B.C. affected the Battle of Syracuse. Both the Carthaginians and the Greeks had settled parts of the south coast of Sicily, resulting in conflicts from time to time. As part of the Peloponnesian War between Athens and Sparta, skirmishes took place far afield, and the Athenians had a major force stationed near Syracuse, ready to move on the offensive. Just then the eclipse was seen and taken to be an unlucky omen. With advice from his soothsayers, the commander, Nicias, delayed departure for almost a month, handing the enemy an advantage. The upshot was that the Athenians were heavily defeated, and Nicias was killed in the fight.

Now step forward to the ninth century A.D. On the first day of that century Charlemagne was crowned emperor of what was to become the Holy Roman Empire. He died in 814, but between 807 and 810 a peculiar set of solar and lunar eclipses had been visible from his kingdom, and their natural cause was explained to him. The trouble started with his son and successor, the first in the very long line of kings of France called Louis. It seems that Louis associated his father's demise with the preceding eclipses, interpreting them as ill-starred portents. When a total solar eclipse occurred on May 5, 840, Louis imagined that the finger was being pointed at *him*. He took fright and never recovered, believing that his days must be numbered. Sure enough, he died a month later. In the aftermath of his early death there was much warring between his three sons, all claimants to the throne. This resulted in the division of much of Charlemagne's empire into the areas we now know as France, Italy, and Germany.

Jumping ahead 800 years, by the seventeenth century both eclipses and comets commonly were held to be signs of awful things to come. This pervading gloomy belief shows itself in the writings of the sages of the day. Consider three of the literary giants of the era. First, William Shakespeare in *King Lear*:

These late Eclipses in the Sun and Moon
Portend no good to us.

Next, John Milton in *Paradise Lost*:

As when the Sun new risen
Looks through the horizontal misty air
Shorn of his beams, or from behind the Moon
In dim eclipse disastrous twilight sheds
On half the nations, and with fear of change
Perplexes monarchs.

Thirdly, poet Samuel Butler thought that a remarkable man was one who could envision the future without making use of eclipses and comets:

> He could foretell whatsoever was
> By consequence to come to pass.
> As Death of Great Men, Alterations,
> Diseases, Battles, Inundations.
> All this without the Eclipse of Sun,
> Or dreadful comet, he hath done
> By inward Light, a way as good.

There is no doubt, then, that there was a common view that eclipses were unlucky phenomena, even deadly. Nowadays an eclipse may be greeted as a great opportunity, but for the greater part of our history they have been subjects of fear and terror. Obviously being able to predict their occurrences well ahead of time would have been a valuable tool.

SUPERSTITIOUS NONSENSE?

We may scorn superstitions such as eclipses, but few people are not afflicted by some irrational belief. The atheist may gesture at religion as a case in point. A baseball player may always put on his left sock first, or insist on being the last out of the tunnel onto the diamond. Professors of logic may avoid walking under ladders or look askance at black cats. Recognizing that superstitions exist can lead to personal advantages: if you must enter lotteries, choose among your numbers 13 and multiples thereof, because relatively few others will do so, and so you would not have to split any prize you won.

The same is true of eclipses. They will bring you luck, either good or bad, if your personal belief system veers in either direction, or if you understand the superstitions of others and act accordingly.

One oft-told tale is of a pair of Chinese astronomers who were brought bad luck by an eclipse. No one is quite sure which ancient society should be accorded recognition as having provided our oldest eclipse record. There are several possible claims for Babylonian and Hindu observations between 1400 and 1200 B.C., but if the story of the Chinese astronomers Hsi and Ho is based on fact then old Cathay possesses the earliest instance. This pair were joint royal astronomers, but they spent too much time studying alcohol and not enough following the Sun and Moon. Solar eclipses were imagined to be the result of the Sun being devoured by a dragon. It was thought necessary to know about such an event in advance so as to organize teams of people to beat drums, yell, and shoot arrows into the air, such a commotion reckoned essential to driving off the dragon. The inebriated duo failed to predict the eclipse, and the emperor was much displeased. Hsi and Ho were even less happy with the outcome: they lost their heads. Our back-calculations show that in the epoch in question, the twenty-second century B.C., several eclipse tracks crossed China, making at least the core of the tale feasible. The favored date is October 22, 2137 B.C., but we cannot be certain on such flimsy evidence.

One might imagine that such a superstitious belief—that eclipses are signs of divine displeasure—must surely be a thing of the distant past. But that is not the case. The first eclipse of the third millennium came eight days after its proper dawning, on January 9, 2001. This was a lunar eclipse that was total as viewed from most of Asia, Africa, Europe, and the eastern seaboard of North America. In Nigeria, much of which has recently come under the influence of Islamic fundamentalists, the eclipse caused great consternation, its advent blamed on sinners. In the northeast of the country there were a rampages by gangs of youths, with more than 40 hotels and drinking houses burnt down in the city of Maiduguri. Similar destruction took place in other towns. "The immoral acts committed in these places are responsible for this eclipse," explained one of the leaders of the riots.

Five months later, on June 21, 2001, the first total solar eclipse of the new millennium was also witnessed in Africa, but further to the south. As the track passed over Angola, Zambia, Zimbabwe, Mozambique, and finally the southern part of the island of Madagascar, thousands of international tourists watched the spectacle they had traveled so far to see. Likewise many millions of local inhabitants gazed skywards in awe, having been briefed by their governments to expect this natural event and warned not to look directly at the exposed Sun. Others, however, were not so confident of the outcome. Some huddled in their mud huts with doors and windows tightly barred, convinced that any light glimpsed from the eclipsed Sun would strike them blind. Elsewhere much wailing and gnashing of teeth accompanied what was regarded as the "rotting of the Sun," from which the world would never recover. It did, of course, quite promptly.

That is not to say that superstitions regarding eclipses are restricted to such places. Take a look sometime at the astrology columns in sundry daily newspapers and weekly magazines, avidly consumed by many millions of readers in our "scientific" Western countries.

THE HUMAN SIGNIFICANCE OF ECLIPSES

Eclipses may have been viewed as being either propitious or portentous by civilizations past and present, but they have influenced our development beyond affecting the outcomes of battles or the deaths of monarchs. Everyone is familiar with the concept of the Sabbath, the one day of rest in seven, taken on disparate days by different religions (Sunday for most Christians, Saturday for Jews and Seventh-Day Adventists, Friday for Muslims). The sabbatical leave (one year in seven, traditionally) of academic staff at older universities is another example.

This is a later adaptation of the original meaning of the Babylonian word *sabattu*, which was considered to be the "evil day" of the moon goddess Ishtar, a time when she was thought to be menstruating, at full moon. This may have come about because of the aforementioned fact that during a lunar eclipse—which can only occur at full moon—the Moon's disk takes on a blood-red hue. Thus the original meaning of the Sabbath was full moon, a monthly rather than weekly event.

It was much later that the seven-day week developed, through a reinforcement during the Jewish Exile in Babylonia (in the sixth century B.C.) between the astrological seven-day cycle employed there and the Judaic Sabbath cycle, in its present meaning. That's where our week comes from: it started out as an eclipse myth.

This provides one example of how eclipses have affected our timekeeping systems. We will see later that the design of our calendar also derives largely from eclipses. It was eclipse records that provided the yardsticks against which the length of the year could be reckoned to a precision of a few minutes, more than a millennium before the construction of the first mechanical clocks.

THE SHAPE OF THE EARTH

Notwithstanding the claims of your local Flat Earth Society, it is a well-established fact that the Earth is round. One might ask, though, when this realization came about.

We credit the idea that the Earth and other planets orbit the Sun to the medieval Polish astronomer Nicolaus Copernicus. It took some time thereafter for various churches to accept that our planet is not a stationary center to the universe. As early as 270 B.C., however, Aristarchus of Samos had suggested that the Earth and other planets circuit the Sun, and to him it was clear that we inhabit a spherical body moving through space. Despite his thinking, it was the cosmology of the second-century A.D. Greek astronomer Ptolemy, with the planets, Sun, and stars circuiting the Earth on convoluted paths, was to hold sway until Copernicus showed the way ahead 1,400 years later.

Leaving the orbits of the celestial bodies aside, the shape of the Earth provided a problem that has an obvious solution (the view from a mountaintop shows its curvature), and yet was much argued about. Eclipses were central to the debates of Pythagoras, Aristotle, and the other Greek philosophers on this question. If the Moon were eclipsed when it passed into the shadow of the Earth, then the shape of that shadow must represent the profile of the planet. Figure 1-11 shows an amusing representation of the argument: if the Earth were square or flat, with edges that sailors could fall off perhaps, then the shadow edges should be flat. They aren't.

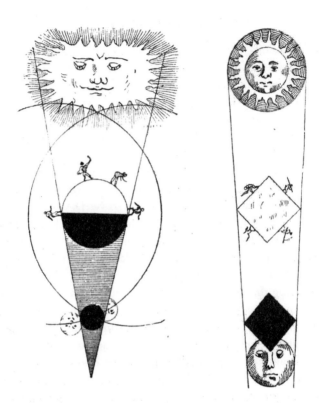

FIGURE 1-11. Lunar eclipses show that the Earth is round. If it were square, then the shadow cast onto the Moon's surface would be the same shape.

ECLIPSE ETYMOLOGY

Before leaving Greece, we should note that the word *eclipse* has a Greek origin. In that language *ekleipsis* means to "leave out," "forsake," or "fail to appear." As with many scientific terms, the less-cultured Romans adopted the Greek word, the Latin becoming *eclipsis*. That word (or the variant *ellipsis*) is directly used in English to imply a place where something is missed out, such as when a printer employs either a dash or three dots in a row. Our word *eclipse*, which can be used as either a noun or a verb, has gone through various spellings in English since about 1300. Variants include eclips, esclepis, enclips, eclypse, and eceps.

An astronomical term that is extensively employed is *ecliptic*, referring to the apparent path of the Sun across the sky. It gets its name because that is where eclipses occur: the Moon must be crossing the ecliptic if it is to line up with the Sun, either in front of or behind the Earth. This term has also been in use for many centuries; for example, in 1391 Chaucer wrote about "the Ecliptik lyne." The word may also be used to refer to the *plane* of the terrestrial orbit.

CHASING ECLIPSES

Random locations on the Earth's surface are traversed by a total solar eclipse track about once per four centuries, on average. Such figures prompt eclipse enthusiasts to travel far and wide in pursuit of their few minutes of heavenly pleasure. With relatively cheap jet travel available, some have experienced totality in a dozen exotic locations, or more. Even seven decades ago the eclipse bug had infected many people, such as Rebecca R. Joslin writing in

1929: "Now eclipses are elusive and provoking things . . . visiting the same locality only once in centuries. Consequently, it will not do to sit down quietly at home and wait for one to come, but a person must be up and doing and on the chase!"

In any distribution there are usually wide deviations from the average, like professional basketball players being taller than the norm, and football linebackers heavier. The path of the total solar eclipse of August 11, 1999 passed from the northwestern Atlantic across central Europe, the Middle East, and then India. On March 29, 2006, a similar event (actually with a wider track and duration of totality) will occur, the path beginning in northeastern Brazil, crossing the Atlantic and then Africa, heading northeast to pass over central Asia, and finishing just short of Mongolia. Those paths have to meet somewhere, and the lucky location is close to the Black Sea coast of central Turkey. No doubt hotel owners and tourist agencies there rubbed their hands in glee when they discovered this: two total eclipses within seven years!

Eclipse chasing for amusement is not a new phenomenon, as seen in Figure 1-12, which captures enthusiasts in Spain in 1900, testing their filters and cameras shortly before totality began. (Spain did rather well, eclipse-wise, around that epoch, with tracks crossing the Iberian Peninsula in 1842, 1860, 1870, 1900, and 1905.) The importance of not looking directly towards a solar eclipse with unshielded eyes had long been recognized by then, and various filters were used, or optical devices like telescopes or binoculars employed to project an image onto a screen as in Figure 1-13.

The use of eclipses by professional astronomers had begun somewhat earlier. By the late nineteenth century, teams from observatories in the developed world were conducting expeditions

FIGURE 1-12. Well-prepared eclipse enthusiasts await the total solar eclipse in Spain in 1900.

in chase of total solar eclipses to the far-flung corners of the globe, not all of them terribly hospitable. In the first decades of the twentieth century astronomers several times trekked to the Far East to observe eclipses, often with unfavorable weather. Figure 1-14 shows the equipment set up under the palms on Flint Island, in the Coral Sea to the east of New Guinea, where operations were hampered by giant land crabs that tried to make off with anything edible, including the astronomers' boots.

WHAT COMES NEXT

Before we can move on, we need to develop an understanding of how eclipses work. Eclipse predictions were made millennia ago,

FIGURE 1-13. How to avoid blinding yourself while trying to observe the Sun with a telescope. This nineteenth-century drawing shows how to project an image onto a screen, an invaluable technique for eclipse watchers.

but a full mathematical model for the orbit of the Moon requires 1,500 separate terms, a quite recent attainment of the science of celestial mechanics. We are able to ascertain in advance accurate eclipse paths only by using high-powered computer codes. (Just as well, else observers would not know where to head for, bearing

FIGURE 1-14. An eclipse expedition in 1901 to Flint Island in the Coral Sea provided a most uncomfortable experience for astronomers; huge land crabs disrupted their preparations.

their telescopes and other paraphernalia.) That poses a bit of a puzzle. How, say, might the Romans have known in advance about the lunar eclipse in 168 B.C., used to their advantage in defeating the Greeks at Pydna? The answer lies with cycles. Eclipses, both solar and lunar, occur in cycles that may be recognized from long-term records. Once one understands the cycles, eclipse prediction is easy.

To unveil the cycles, one could patiently record eclipses for the next several decades. A more sensible approach would be to look up the available information on the dates, times, and locations of eclipses over the past few centuries and then try to decipher the code. That is just what the Babylonians, Greeks, and others did more than two millennia ago, so it is not impossible, but it is tedious.

Instead one could tackle the project backwards. Starting with our modern knowledge of the Moon's orbit about the Earth, and the Earth's orbit about the Sun, we may deduce when eclipses will occur, and with what sorts of repetition cycles. Remember we are way ahead of the ancients, who thought that the rest of the universe revolved around the Earth. We know about orbits and how to make the relevant calculations. Such analyses can be explored in detail in the Appendix.

In Chapter 2 we will consider the orbits of the Moon and the Earth and show how our calendar depends upon them, before looking at some general features of eclipses and their cycles. Then in Chapter 3 we will delve deeper into the history of eclipses, solar and lunar, and explore their significance in the chaotic course of civilization.

2

The Heavenly Cycles

And the Moon in haste eclipsed her
And the Sun in anger swore
He would curl his wick within him
And give light to you no more.

Aristophanes, *The Chorus of Clouds*

Our life rhythms are controlled by the heavenly cycles: the daily rising and setting of the Sun, the monthly variation in the brightness of the Moon, and the seasonal north–south displacement of the Sun affecting the influx of solar power and thus the climate.

It does not necessarily follow that all plants and animals have identical tempos. For any cyclic phenomenon, scientists may speak of both its *frequency* (the number of times it occurs within a given time, or equivalently its *period*, the duration of the cycle), and also its *phase*. Football games, for example, are played once a week (the frequency), making the period seven days, but there are different phases depending upon the level involved: high schools tend to play on Friday evenings, college matches are on Saturday afternoons, and professional games on Sundays.

In the natural world most organisms follow a basic daily cycle, although their phases may differ. The majority of animals go about their business during daylight hours, but there are many specialized nocturnal beasts, too. (To go back to our football metaphor, there are also a few games on Monday or Thursday nights.) In addition, not all animals follow 24-hour cycles, as we will see.

Turning to the yearly cycle, the changing levels of daylight and temperature influence us all, and more so at extreme latitudes rather than the tropical zones where intra-annual variations are minimized. Who can claim that they are not affected in *any* way by the seasons, if only through oscillations in the price of fresh foods? Other animals suffer more radical alterations in food availability during the year, which exert greater control over their lives. In consequence many species hibernate during the winter, emerging only when the signs of spring promise plenty of food, telling them it is time to eat and breed again.

THE CALENDRICAL INFLUENCE OF THE MOON

The Moon also affects us in sundry ways. Much has been made by numerous authors of the apparent fact that the human female menstrual cycle has an average duration of about 29.5 days, which is indistinguishable from the cycle of lunar phases. One might consider this to be a coincidence. On the other hand it might be evidence of a causal relationship, perhaps amenable to scientific analysis; for example it may be that repeated high fertility near new moon, when it is dark and dangerous to roam at night, favored reproductive success in early humans.

Quite apart from this physiological cycle, it is indisputable that our natural satellite exerts control over our affairs. First, we should look at the calendar (or perhaps I should write *calendars*, because different nations and religions use calendars other than the familiar calendar to which we in the Western world are habituated). In the Western calendar the months, despite the etymology of that word, are no longer linked to the lunar phases, which is why I differentiate between lunar and calendar months. That divorce between the calendar month and the Moon, in the evolutionary history of the Western calendar, occurred before 400 B.C., in the Roman republican era before Julius and Augustus Caesar instigated Imperial Rome. The influence of learned Egyptians upon Julius Caesar ensured that his reformed calendar post-46 B.C. was an exclusively *solar* calendar.

Other calendars retain the influence of the Moon. While the civil Western calendar is solar, the ecclesiastical calendar endorsed by Pope Gregory XIII in A.D. 1582 (called the Gregorian calendar) is *luni-solar*. That is, the Moon defines the dates of all moveable feasts in the liturgical year, reckoned from Easter, which is based on the full moon after the spring, or vernal, equinox. (Beware, though, that the days of the full moon and the equinox used to derive the date of Easter Sunday are founded upon ecclesiastical rules, rather than the real Moon and Sun in the sky.) The Hebrew calendar is also luni-solar, with the Moon affecting the dates on the Western calendar for Passover, Rosh Hashanah, and Hanukkah. Similarly Chinese New Year is often celebrated at the second full moon after the winter solstice, although again the precise rules are complicated. In such calendars the Sun still defines the average (mean) length of the year, because the holidays are simply phased according to the lunar cycle following some solar-defined juncture (one of the equinoxes or solstices).

The Islamic calendar contrasts with this, the year being defined *exclusively* by the Moon, the annual round containing 12 lunar cycles. Because each month starts with an observed new moon, and there is only one chance a day to witness this (after sundown), the months and the years must contain a discrete number of days. On average, the Islamic year lasts for 354.37 days, but particular years generally contain either 354 or 355 days, and more extreme lengths are possible because of vagaries in spotting the new moon owing to atmospheric conditions. The Islamic lunar year is thus 10 or 11 days short of a solar year, and the calendar slips through the seasons on a cycle of almost 34 years.

More information about the astronomical bases of different calendars is given in the Appendix.

THE TIDAL INFLUENCE OF THE MOON

The Moon also influences us through the tides, which are raised by the attraction of the lunar gravity on our oceans. While the Sun also plays a role, resulting in the contrasting heights attained by spring and neap tides, the major cause is the lunar attraction. At the side of the Earth nearest the Moon the oceans bulge upwards due to its pull. On the far side of our planet the seas also bulge outwards *away* from the Moon's direction (in simple terms this is because that part of the globe is furthest from the Moon, its gravitational pull minimized there).

The tides do not follow a 24-hour cycle. This is because, during the time the Earth takes to spin on its axis, the Moon has moved some distance along its orbit around us. The latter body does not return to the same place in the sky until 24 hours and 50 minutes later; this is a whole day plus one part in 29.5 (the number of days the Moon takes to orbit the Earth). The effect is that the times of high tides shift progressively later by almost an hour each day.

To someone in the developed world interested in boating or fishing this may merely prove a nuisance, but to maritime societies such as in Greenland, or the Melanesian and Polynesian islands in the Pacific, the tide timetable is fundamental to their livelihood. Thus their "day" would not be based on the movement of the Sun, a 24-hour cycle, but rather a lunar day, lasting 24 hours and 50 minutes.

In the natural world many animals living in mangrove swamps and intertidal mudflats are similarly affected by the Moon. Their daily routine follows not the cycle of sunlight, but rather the rhythm of the tides, which is controlled by the spin of the Earth and the orbit of the Moon.

Clearly the Moon affects both the human and the natural world in diverse ways. It is not just some lifeless lump of rock forever circuiting our planetary home as a mere curiosity. We have good cause to want to understand its cycles, which are both complex and remarkable. The length of the year we use in religious calendars and so forth may be directly affected by the presence of the Moon. Long-term changes in the dates of the solstices and equinoxes are caused mainly by tugs imposed on the orientation of the Earth's spin axis by the Moon. The Moon's various cycles are considered next in further detail, along with their effect on the pattern of eclipses.

THE CYCLES OF THE MOON

As we have already seen, our Western calendar months are divorced from the Moon. Let us leave them aside. There are several astronomical definitions for the month, each taking some specific phenomenon as its basis. These are each of vital significance in determining eclipse cycles, and they are all defined and discussed in the Appendix. Immediately, however, it is the *brightness* cycle of the Moon that we want to know about, and so the length of the month appropriate is that from one full moon to the next. That is called the *synodic month*, or sometimes a *lunation*, or simply *lunar month* (the latter terms perhaps being somewhat ambiguous). It lasts for about 29.5 days, as we mentioned previously. Any particular synodic month may range in duration by six or seven hours from the mean (between about 29.2 and 29.8 days), the average over several years being 29.53059 days.

When the Moon is aligned with the Sun we say it is at *conjunction*, whereas when it is 180 degrees from that point it is at *opposition*. (Although opposition is the time of full moon, strictly speaking conjunction is *not* the time of new moon. This is because for the new moon to be visible near the western horizon just after sunset, it needs to have moved along its orbit so it is sufficiently separated from the Sun. Conjunction may be thought of as being "dark of moon," when it cannot be seen at all in the solar glare.) The Moon may be said to be in *syzygy* when it is at either of these points. Eclipses can occur only near syzygy.

Until now we have been effectively assuming that the Earth, Sun, and Moon all inhabit the same plane. If this were the case then the Moon would cross the face of the Sun, producing an eclipse, every time it passed conjunction. The lunar orbit does not remain in the same plane as that which the Earth occupies, a matter of great importance with regard to eclipses. If we consider the Sun and the Earth's orbit around it to be in the plane of the paper in this book, then the Moon's orbit is, in reality, tilted by about five degrees to that plane; this angle is called the *inclination*.

The Sun has a diameter about 109 times that of the Earth, while the Moon is not much more than a quarter the size of our planet. This means that to get an eclipse requires a quite stringent alignment. An eclipse occurs *only* if the Moon crosses the ecliptic when very close to either conjunction or opposition, respectively producing solar and lunar eclipses. During each circuit of the Earth, the Moon crosses the ecliptic once traveling upwards, and once traveling downwards. These are called the *nodes* of the orbit, the *ascending* and *descending* nodes respectively. An eclipse can occur only at a node.

An angle measured counter-clockwise around the ecliptic from the location of the spring equinox is called a celestial *longitude*, a similar parameter to the geographical longitudes we use to produce a grid on the Earth's surface, the Greenwich meridian being the fundamental reference. Astronomers likewise use celestial *latitudes* for the angle north or south of the ecliptic plane.

THE METONIC CYCLE

You may recall that, at the beginning of the book, we mentioned the period of 235 synodic months, which lasts for almost exactly 19 years. The difference between these is just 125 minutes. Astronomers call this 19-year period the *Metonic cycle*, after the mathematician Meton who lived in Athens in the fifth century B.C., although there is evidence that the Babylonians knew of the synchrony earlier.

Meton invented a calendar cycle containing 6,940 days, but the Greeks never adopted it for widespread use. The 19-year cycle *is* employed, though, in various other calendrical spheres. Many past and present calendar schemes using leap *months* rather than the familiar leap *days* have seven extra lunar-based months spread over 19 years. Thus a dozen of the years each contain 12 months, while seven of them have 13, making 235 in all.

A form of the Metonic cycle is used today in the Hebrew calendar, and many other luni-solar calendars. It is also employed in the calculation of Easter. Its inaccuracy—that discrepancy of 125 minutes—has affected history in various ways, most especially in the evolution of ecclesiastical calendars. Again the intricate details are discussed in the Appendix.

The discovery of the Metonic cycle by the ancients would have been possible simply by watching for a repeated full or new moon at the same time of year. The fact that there are almost exactly 235 lunations in 19 years is a phenomenon that could have been identified by quite early societies, such as those who were building megalithic monuments in Britain and elsewhere in Western Europe from at least the middle of the fourth millennium B.C. This simple coincidence between the lunar and solar cycles would have been fairly impressive, and most important would have allowed the subsequent prophecy of various celestial events. Starting from that basis further coincidences would soon be unveiled. Such considerations may underlie the gradual evolution of impressive sites such as Stonehenge.

THE ECLIPSE YEAR

If the lunar orbit were fixed in space, such that the nodes occurred always in the same locations, then the Sun would pass through those nodes once per solar year. This is not the case, though; in fact the nodes are moving backwards around the lunar orbit, a motion that is termed *precession*. (This is similar to the way in which a toy gyroscope twists around.) In consequence the Sun gets to the nodes (where eclipses may occur) progressively earlier, producing a type of year that is somewhat shorter. This is called an *eclipse year*, lasting about 346.6 days, and it is a crucial cycle of time because it will affect the frequencies and characteristics of eclipses. In any calendar year one will usually find pairs of solar eclipses separated in time by close to half an eclipse year (173 days). An example is June 10 and December 4, 2002, a separation of 176 days.

Why is this slightly more than half an eclipse year? There are two contributing factors. First, the Sun does not move across the sky at a constant rate throughout the year, because our orbit is not precisely circular. The Sun's apparent speed is slowest around June/July, when it is furthest from Earth (astronomers call this *aphelion*), and quickest around December/January, when we are closest to that orb (*perihelion*). Second, and more significant, eclipses do not necessarily occur precisely *on* the node, but rather there is a range of possible positions called the *ecliptic limits*. These limits are defined in the Appendix.

During each eclipse year, eclipses can take place only while the Sun and Moon are within the ecliptic limits, defining periods known as the *eclipse seasons*. The lengths of such seasons depend upon the eclipse type in question: Solar or lunar? Are they partial or total? We will discuss this matter shortly and show how the eclipse seasons come about in the Appendix.

THE SAROS

There is also a long-term cycle over which conjunctions and oppositions repeat, making eclipses possible. This period is known as the *saros*, a Greek word meaning "repetition" that is itself derived from the Babylonian *sharu*. In eclipse calculations the saros is of huge importance.

On our calendar we will see eclipses repeating with gaps of 18 years plus 10 or 11 days (a length of time very close to 19 eclipse years, or 18.03 solar years). Take, for example, the eclipse of August 11, 1999. It was preceded by a similar event on July 31, 1981, and will be followed by another on August 21, 2017. The first saros gap had four leap years (1984, 1988, 1992, 1996) so each date within the year was 11 days earlier, while the second saros gap contains five leap years (2000, 2004, 2008, 2012, 2016) leading to the next date being but 10 days later. Knowledge of this cycle (the *saronic cycle*) therefore allows a sky watcher or astronomer to make long-term eclipse predictions.

At any time there are several distinct saronic cycles in action, interwoven but distinguishable. By now you will have got the picture that due to a host of coincidences of celestial mechanics, there are various underlying cycles that make eclipses repeat in a rather predictable way.

TYPES OF ECLIPSE

A scientific understanding of any phenomenon starts by sorting the available observations into appropriate groupings based upon some fundamental characteristic. We sort small six-legged beasts into the category "insects," while eight-legged ones are called "arachnids" (spiders, scorpions, mites, ticks, etc.). Naturally other considerations also apply: an octopus is not an arachnid.

Similarly the basic eclipse phenomenon is subdivided into different types. The first distinction, as we have already seen in Chapter 1, is between lunar and solar eclipses. Up to this juncture we have been concerned mainly with *solar eclipses*, produced by the Moon passing between us and the Sun. Similarly, the Earth may circulate between the Sun and the Moon, putting the latter into its shadow. This is a *lunar eclipse*.

A second distinction is between total and partial eclipses. In a *partial eclipse* the alignment of Earth, Moon, and Sun is not exact, so only part of the disk of the Sun (in a solar eclipse) or Moon (in a lunar eclipse) is obscured.

Considering now only solar events, one can get a perfect alignment, but nevertheless not a total eclipse. This is due to the mutual separations of the three bodies varying. As the Moon intercedes between the Earth and the Sun, it may or may not be large enough to block out the whole of the solar disk. This is because its orbit is not circular, so it is sometimes closer to the Earth (at *perigee*) and sometimes further away (at *apogee*). When near perigee its disk appears comparatively large (refer to Figure 1–2) and so can cover the Sun completely—a *total eclipse*. When it is at apogee, its disk appears smaller and so it is unable to obscure the Sun completely. A bright "annulus" then appears around the circumference of the Moon, and so this is called an *annular eclipse*. These three situations are depicted in Figure 2-1.

There is a fourth type known as a *grazing eclipse*. It occurs when the limb of the Moon just touches the apparent edge of the Sun in the sky, but does not overlap it. Solar eclipses may also be of

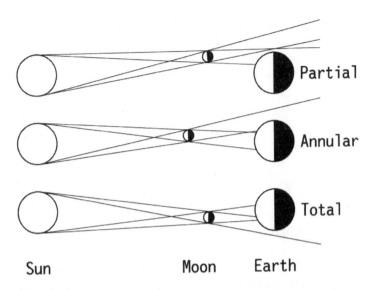

FIGURE 2-1. Three basic forms of solar eclipse can occur. If the Moon does not pass centrally over the solar disk, then the eclipse is only partial. If the passage is close to central, then the eclipse may be either total or annular, depending on the distance from us of the Moon and, to a lesser extent, the Sun. A total eclipse occurs if the rays from top and bottom of the Sun touching top and bottom of the Moon do not cross before reaching the Earth. If those rays do cross above the Earth's surface, then the eclipse will be annular, with a ring of the solar surface being visible around the Moon's periphery. In this diagram the sizes of the three bodies are highly exaggerated.

hybrid nature, total in some locations and annular in others. All these situations are explained in more detail in the Appendix, including the additional influence of the Earth-Sun distance varying during the year.

GEOGRAPHICAL SHIFTS IN ECLIPSE PATHS

Apart from regularly repeating in time according to the saros, another effect associated with that 18.03-year cycle in eclipses is a consistent shift in the geographical location of the track of totality. There is both a small step in latitude (the north–south direction) and a larger shift in geographical longitude (the east–west direction). Let us examine the origin of these shifts.

The saros actually lasts for 6,585.32 days. Knocking off the integer number of days, there is an excess of just less than one-third of a day, representing almost one-third of a rotation of the planet. In terms of *time*, it is equivalent to 7 hours and 41 minutes; in terms of geographical longitude, it means that the eclipse track is shifted by about 115 degrees to the west from one saros to the next.

The latitudinal offset occurs because after one saros has elapsed the Moon takes a slightly different path across the face of the Sun. This is explained in more detail in the Appendix. From one eclipse to the next in any saronic cycle, the step in latitude is about four degrees. This may either be northwards, or southwards, depending on whether the Moon is at its descending or ascending node (passing through the ecliptic moving either south or north). These two effects (north–south and east–west shifts of eclipse tracks) are illustrated in Figure 2-2, with six twentieth-century total solar eclipses in a sequence known as saros 136 delineated. After each 18.03-year gap there is a consistent offset in the path of totality: westwards by 115 degrees, northwards by 4 degrees.

Although the *tracks* move from east to west from one saronic cycle to the next, the actual path followed by the lunar shadow traces across the Earth from west to east, because the Moon is

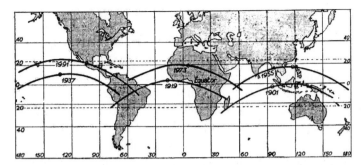

FIGURE 2-2. The six long-duration eclipses of the twentieth century were all members of the same saronic cycle. After each gap of 18.03 years, another seven-minute eclipse occurred, displaced westwards by 115 degrees and northwards by 4 degrees. From the onset of each of these tracks to its end took about five hours; that is, totality occurred at quite different times in separated locations.

overtaking the Sun in the sky. For example, the eclipse of 1991 shown in Figure 2-2 started to the southwest of Hawaii, crossed the eastern Pacific, passed over Mexico and other parts of Central America, and finished over Brazil.

I have just mentioned that these were all total eclipses, which might seem unexpected: would not a mix of total, annular, and partial eclipses be anticipated? The answer is *NO*. The reasons for this are explored in the Appendix, but the pertinent point here is that the basic characteristics of the six eclipses in Figure 2-2 repeated; they did *not* comprise a random hotchpotch of partial, annular, and total eclipses. Indeed this is a particularly prominent sequence as they had the longest periods of totality (six or seven minutes) of any solar eclipses in recent centuries. Obviously something systematic happened, and this is another fundamental quality of any saros sequence. Again, we delve into this in the Appendix.

A final note on the sequence shown in Figure 2-2: it is not finished yet, with the next members being due on July 22, 2009 and August 2, 2027, each lasting for about six and a half minutes (the lengths are decreasing from a peak in 1955). It should be easy enough to extrapolate from that diagram and work out the eclipse tracks in those years, in case you want to make travel plans: the 2009 path will cross eastern Asia, while in 2027 northern Africa will be the place to be.

THE SHAPES OF THE ORBS

So far it has been assumed that the Earth, Moon, and Sun are spherical. In reality, because of their rotational properties they are each slightly flattened into shapes known as "oblate spheroids." The Earth is actually somewhat pear-shaped, the northern hemisphere a little thinner than the south, and both our planet and the Moon are also a little rough around the edges (both possess mountains and so on). At least their shapes are constant in the short term, whereas the Sun is forever throwing out material in coronal loops and prominences.

For simplicity, however, we assume spherical profiles in the following text, and thus circular disks and shadows. We turn next to the characters of these shadows.

THE UMBRA AND PENUMBRA

The shadow cast by the Moon onto the terrestrial surface has a form as sketched in Figure 2-3. The dark central spot is the region

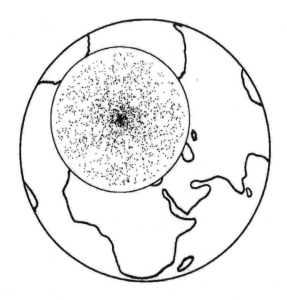

FIGURE 2-3. The umbra (complete shadow) of the Moon on the Earth is a solitary black spot in the middle of the pattern shown here. It is quite small, typically only 60 miles across, whereas the partial shadow or penumbra is over 4,000 miles in diameter, covering a large fraction of the dayside during an eclipse.

of totality: the Moon as seen from anywhere within that small region completely covers the solar disk, and this totally shadowed spot is called the *umbra*. Typically the umbra, or the path of totality, is 60–100 miles wide, although it can have effectively zero width (as in the case of a transition between a total and an annular eclipse), or be intrinsically a little wider in the longer eclipses. There is also a geographical effect: the lunar shadow may be cast

FIGURE 2-4. During the total solar eclipse of August 11, 1999, the crew of the Russian space station Mir photographed the Moon's shadow moving across the Earth's surface.

obliquely onto the Earth's surface, and so the width of the ground track tends to be wider for eclipses in the Arctic or Antarctic.

All around the region of totality the Moon only partially obscures the Sun, and this partial shadow is termed the *penumbra*. The penumbra is much wider than the umbra. While the umbral spot may have a radius of only tens of miles, the penumbral radius is 2,000–2,200 miles. A partial eclipse will be detectable anywhere within that large area, a grazing touch between lunar and solar disks occurring at its very edge.

Figure 2-4 is a photograph of the Earth obtained looking down from orbit during an eclipse, showing the lunar shadow very clearly. It is the people under this dark central spot who experience totality.

If you were living in Babylon a few thousand years ago, only a tiny fraction of all total solar eclipse paths would cross that city. The penumbra for a total eclipse seen elsewhere would cross Babylon in about a quarter of all cases, because the penumbral circle in Figure 2-3 scans about half of the dayside face of the globe (and half the time you would be on the night side). Mostly the city would lie towards the periphery of that shadow and the Moon would only cover perhaps 10 or 20 percent of the Sun, so that the eclipse might well be missed without foreknowledge. For a society constrained to Mesopotamia and environs, only a small fraction of all solar eclipses would appear in the records, making the discovery of the complex cycles described earlier a near-impossibility.

How, then, were the eclipse cycles unveiled?

LUNAR ECLIPSES

That question may be answered by considering lunar eclipses. First, note that the frequency of lunar eclipses is not the same as the frequency of solar eclipses. Although the Earth is bigger than the Moon, so that it casts a larger shadow, the Moon is a smaller target for that shadow to hit and so, overall, lunar eclipses are not so numerous. There are on average 238 solar eclipses per century, but only 154 lunar eclipses.

Despite their comparative infrequency, for an observer restricted to one position on the terrestrial surface (say, an ancient Babylonian astronomer), lunar eclipses are witnessed more often than solar. This is because the full moon may be seen from *anywhere* on the night side of the planet. That implies that half of humanity might see the Moon being eclipsed, but in addition such eclipses last several hours, and the globe spins to allow observers elsewhere a chance to note the eclipse, even if all the phenomena may not be seen from the extreme locations.

An example of a lunar eclipse, that of May 16, 2003, is shown in Figure 2-5. Throughout South and Central America, and the Atlantic, the entire eclipse may be witnessed. The start of the eclipse, and all of the total phase, can be seen throughout the contiguous United States and most of Canada. The same is true for Europe and Africa. In the western parts of North America the Moon will be in eclipse as it rises. Europe will miss the final stages of the eclipse because the Moon sets during the process.

THE SELENELION OR HORIZONTAL LUNAR ECLIPSE

If you are interested in experiencing something that very few other eclipse watchers have seen, this event in 2003 may provide you with an opportunity. If you were in one of those zones where the eclipse is in progress at moonrise or moonset you have the peculiar chance to be able to see both the Sun and the eclipsed Moon in the sky at the same time, with a quick twist of the head. An eclipse occurs when the two celestial orbs are 180 degrees apart, with the Earth in between. The refraction (or bending) of light beams in the Earth's atmosphere, however, makes it possible to see both at once. Geometrically they may both be below the horizon, but the refraction by about half a degree makes this double appearance possible. In order to witness this you need to go to as high an altitude as possible and have a clear distant horizon to both the east and west. You have only a fleeting chance lasting a few minutes.

In general this is possible only with the partially eclipsed

Total Lunar Eclipse of 2003 May 16

| Geocentric Conjunction = 03:24:59.1 UT | J.D. = 2452775.642351 |
| Greatest Eclipse = 03:40:08.7 UT | J.D. = 2452775.652879 |

| Penumbral Magnitude = 2.09990 | P. Radius = 1.3118° | Gamma = 0.41213 |
| Umbral Magnitude = 1.13383 | U. Radius = 0.7739° | Axis = 0.42103° |

Saros Series = 121   Member = 55 of 84

Sun at Greatest Eclipse
(Geocentric Coordinates)

R.A. = 03h30m07.2s
Dec. = +18°59'20.2"
S.D. = 00°15'49.2"
H.P. = 00°00'08.7"

Moon at Greatest Eclipse
(Geocentric Coordinates)

R.A. = 15h30m43.0s
Dec. = -18°35'32.4"
S.D. = 00°16'42.2"
H.P. = 01°01'18.1"

N

W

Ecliptic

P1
U1
U2
U3 Greatest
U4
P4

Earth Umbra

Earth Penumbra

S

Eclipse Semi-Durations

Penumbral = 02h34m46s
Umbral = 01h37m20s
Total = 00h26m22s

Eph. = Newcomb/ILE
ΔT = 66.1 s

| 0 | 15 | 30 | 45 | 60 |
Arc-Minutes

F. Espenak, NASA/GSFC - Tue, 1999 Jun 01

Eclipse Contacts

P1 = 01:05:23 UT
U1 = 02:02:49 UT
U2 = 03:13:46 UT
U3 = 04:06:31 UT
U4 = 05:17:28 UT
~4 = 06:14:55 UT

60° N
30° N
Latitude 0°
30° S
60° S

U4 U3 U2 U1 P1   P4 U4 U3 U2 U1 P1   P4
Eclipse at
MoonRise   All Eclipse
Visible   Eclipse at
MoonSet   No Eclipse
Visible

190° W   120° W   60° W   0°   60° E   120° E   190° E
Longitude

Moon, because when the eclipse is total the Moon is simply too dark to see when it is also right on the horizon. At that time you are looking through such a thickness of atmosphere that the weak light from the totally eclipsed orb is attenuated to leave almost nothing. The best chance is when there is still a thin crescent of the lunar disk illuminated by the Sun, meaning between contact points U1 and U2, or between U3 and U4 in Figure 2-5 indicates that the Hawaiian Islands are a candidate location, although many parts of the western United States will also provide an opportunity.

This phenomenon is called a "horizontal eclipse" or, from a French term, a "selenelion." There is evidence that the Babylonians noted an occurrence of this peculiarity in 1713 B.C. In modern times the first record seems to date from 1590, when the great astronomer Tycho Brahe saw a selenelion from his observatory on the island of Ven that lies in the strait between Sweden and Denmark. Five more such events were recorded from Europe over the next century (see Figure 2-6 for an example from 1666), but none in the 1700s and only one during the 1800s. The next record was not until 1975, when Allan Fries noted a selenelion from an island in the sound off Everett, Washington. The first photograph of a selenelion was not obtained until July 16, 1981, when professional

FIGURE 2-5. Details of the total lunar eclipse that will occur on May 16, 2003. Various types of pertinent data are shown, such as the timings for different contact points and the locations from where the eclipse may be viewed. Note the angle between the path of the Moon and the ecliptic: the eclipse occurs near the descending node. Also note that the nodal passage occurs well after the eclipse has finished, an example of the idea of the ecliptic limits.

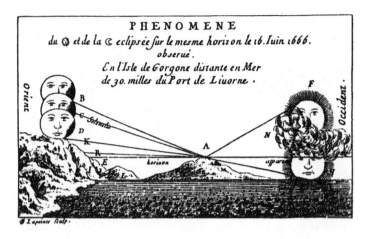

FIGURE 2-6. A sketch of the circumstances of the lunar eclipse observed on June 16, 1666. Prince Leopold of Florence instructed his astronomers to go to the island of Gorgona, 30 miles off the Italian coast near Livorno, in order to record what was seen. The flat Mediterranean Sea provided their horizon to the west where the Sun was setting. By gaining some altitude the distant Appenines were visible low in the east where the Moon was rising, and they were able to witness both the Sun and the eclipsed Moon in the sky at the same time. Such an observation is known as a "selenelion."

astronomer William Sinton permanently recorded one from the Mauna Kea Observatory in Hawaii.

That brings us to the present. A few days before the lunar eclipse on January 9, 2001, I realized that a selenelion might be seen from Adelaide in South Australia, where I had lived for some years. So I alerted friends in the local astronomical society, suggesting that they might try to catch a glimpse from Mount Lofty, the tallest hill on the eastern fringe of the city. I knew that to the west they would look out over the sea, while to the east their view would be over the extensive plains through which the River Murray flows. And in January, at the start of the southern summer, the sky was almost certain to be clear. A small group rose early and climbed not only that hill, but also the fire-spotting tower at the summit, from where they were afforded an excellent view. The result is shown in Figure 2-7.

THE DURATIONS OF ECLIPSES

Total solar eclipses are brief. Although a small fraction last for as long as seven minutes, most present a period of totality lasting only two or three minutes. The partial phase of such an eclipse lasts for much longer, some hours.

Refer back to Figure 2-3, and imagine that you are waiting somewhere on the track that the spot of totality will eventually cross, blanking out the Sun for a couple of minutes. The radius of the footprint delineating the penumbra is about 2,000 miles, and it sweeps across the globe at around 1,600 miles per hour. This means that the partial phase starts about 75 minutes before totality is achieved and continues thereafter for a similar interval. People located well north or south of the track will see only a partial eclipse, but it may last for a couple of hours.

The specifics may be rather different for particular solar eclipses, especially for observers situated close to the edges of the planet in this view, but the broad picture is correct: totality lasts for a couple of minutes, partiality for over an hour before and after.

How long do lunar eclipses last? Since the Moon is large, it is conventional to define several distinct *contact points* or times, as shown in Figure 2-5. The Moon is within the penumbra between

FIGURE 2-7. A selenelion photographed over the city of Adelaide, South Australia, in January 2001. The Sun has just risen in the east, behind the photographer, and its feeble light is starting to illuminate the city, although street lamps can still be seen. The western horizon, out over the sea, can hardly be distinguished in the gloom. Less than a degree above the horizon is the Moon, in partial eclipse, meaning that both the Sun and the eclipsed Moon may be seen at the same time. This is only the twelfth time in history that such an occurrence has been recorded.

P1 and P4, which lasts for up to five-and-a-half hours, during which time the Earth has executed almost a quarter of a revolution. In principle this would allow 70 percent of the planet's inhabitants a chance to see that a lunar eclipse is underway. The umbral stage is much more noticeable. The phase of totality, between U2 and U3, may last for 80 to 90 minutes, but can be much less if the Moon is slightly further north or south compared to the terrestrial shadow. These contact points for lunar eclipses, and the equivalents for solar eclipses, are discussed in more detail in the Appendix.

THE BRIGHTNESS AND COLOR OF THE MOON IN TOTAL ECLIPSE

During a total solar eclipse the Sun's disk gets very dark indeed—you can't see it—but the same is not true of a total lunar eclipse. While it is entirely within the umbra the lunar disk brightness drops to about one part in 5,000 that of the near-full moon, and so it can still be seen. One needs no sophisticated equipment to recognize that the normal bluish-white Moon appears a reddish-brown during the eclipse, and many describe the Moon as taking the color of blood. How does any sunlight at all get to the Moon to provide it with some dim yet red illumination?

The answer lies with the atmosphere. Our planet possesses a considerable atmosphere, and that makes its edge somewhat fuzzy. On the other hand, the Moon has no atmosphere of which to speak, and so it casts a shadow whose sharpness is limited only by the Sun's finite size: when a solar eclipse reaches totality it is sudden and abrupt.

Why does red light preferentially get to the Moon during a total lunar eclipse? This occurs because of inequalities in the atmospheric transmission of different wavelengths of light. This effect actually occurs all the time and is obvious once one thinks about it. At sunset the image you see of our star as it sinks below the western horizon is much redder than at midday because the air molecules between your eyes and the Sun scatter light at the blue end of the spectrum more than at the red end, allowing more of the red light to reach the planet's surface directly, but by the same token making the sky look blue.

The images of the Sun and Moon at rising and setting are also distorted somewhat, producing oval rather than circular profiles. This is due to refraction (that is, bending of light) in the atmosphere; it is similar to the way in which your arm seems to develop a sharp kink when thrust through the surface of a swimming pool. The amount of refraction produced in the atmosphere again depends upon the wavelength of the light in question, just as white light passing through a prism is split into the constituent colors of the rainbow. (It is a fallacy that the Sun and Moon are actually larger in size at rising or setting. This is an illusion produced by having reference objects visible along the horizon, as compared with none when the orbs are overhead.)

The atmosphere can thus produce coloration through two means. One is the fact that the blue end of the spectrum is more efficiently scattered by individual air molecules. The second is that the amount of refraction similarly varies across the spectrum.

At sunset the Sun looks red, but think of the light passing ten miles above your head, skimming through the atmosphere. The blue light is largely being scattered, making the sky blue, and also being refracted to such an extent that it is directed more towards the ground, pushing it deeper into the atmosphere and therefore suffering even more scattering. The red light is more likely to escape scattering and may be refracted by just enough to direct it towards the Moon. What is happening is shown schematically in Figure 2-8.

All around the globe the atmosphere is acting to transmit a

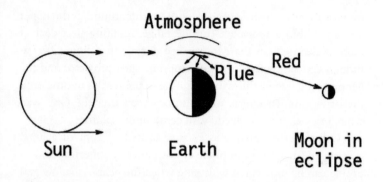

FIGURE 2-8. Why does the Moon turn the color of blood during a total lunar eclipse? Sunlight enters the atmosphere (the thickness of which is shown greatly exaggerated here) and the blue end of the spectrum is preferentially scattered, making the sky blue. This means that more of the red light makes it through on a route to the Moon. In addition, blue light is refracted more (has its course bent) by the air and fails to make it along the necessary direction (not to scale).

little sunlight to the Moon, and that small fraction that makes it through is predominantly at the red end of the spectrum. That is why in a total lunar eclipse the Moon appears a dark reddish-brown. Any dust suspended in the air will add to these effects.

As the curved shadow of the Earth creeps across the Moon, its boundary is blurred, producing a graded fringe rather than a sharp edge. This is because of both the finite solar diameter and the terrestrial atmosphere, but the presence of a substantial atmospheric loading of dust will cause the normal shadow profile to be altered. After major volcanic eruptions, such as those at Mount St. Helens in Washington and Mount Pinatubo in the Philippines, the dust left lofted in the atmosphere may be immense and take months to years to settle out. It tends to drift around in the upper air, but within a restricted latitude range, resulting in a distinct blob of darkness on the Moon's face during an eclipse. It is as though the Sun is acting as the lamp in a slide projector and the Moon as the screen, throwing an image of the Earth's outline onto the latter. Similar effects were also observed after the Gulf War, when oil-well fires produced vast, dense smoke plumes.

Other features may also be investigated by dint of a lunar eclipse. The Earth's atmosphere is not perfectly spherical, and its profile can be monitored by timing when the eclipse shadow gets to various marker points on the lunar surface, such as well-known craters.

It is only during a lunar eclipse that the influence of our atmosphere in these regards is obvious. Lunar eclipses are quite unlike solar eclipses, then.

THE DISTRIBUTION OF ECLIPSES

There are about 66 total solar eclipses per century, but many have ground tracks unfavorable for potential viewers. These may be located either at very high latitudes (over the Arctic or Antarctic), or over regions in which the weather is likely to be poor, such as the tropics during the monsoon, or completely over the ocean. In practice, a total solar eclipse track traversing accessible places with a good chance of clear weather occurs about once every three years. Nevertheless there is many a keen eclipse watcher who has spent an enormous amount of time and money getting to a well-considered prime spot, only to be stymied by an unseasonably cloudy day.

The number of eclipses per century given above is an average, which would result if they occurred randomly in time. The reality, though, as we have seen, is eclipses are not random at all; they repeat on regular cycles. The tracks of total solar eclipses within specific saros sequences advance consistently by steps across the Earth, as in Figure 2-2, and there are other systematic trends.

Eclipses do not occur randomly in terms of geography either. If they happened entirely by chance then any particular location would get a total solar eclipse about once every 410 years. In fact more of these events occur during the summer than the winter in the Northern Hemisphere, because the Earth passes aphelion in July, when the Sun has its smallest apparent size. This makes it more likely that the Moon will be large enough to cover it completely. The Northern Hemisphere is tilted towards the Sun in the summer, meaning that there is a greater probability that it will get an eclipse, and so places north of the equator are visited about once per 330 years. Being the summer there's also a greater chance of clear skies. In contrast Southern Hemisphere locations receive total solar eclipses about once per 540 years. As the bulk of the population lives in the north, this quirk of nature increases the likelihood that a person picked at random from the whole of humankind will experience a total solar eclipse without needing to chase after one.

EXTRAPOLATING ECLIPSES FORWARD

Stepping back some millennia, any civilization in the temperate and tropical zones of the Northern Hemisphere would have been able to register up to 70 percent of all lunar eclipses. Apart from those a lesser number of solar eclipses would have been seen, and the oft-repeated relationships between them noted, such as a solar eclipse often being preceded or succeeded by a lunar eclipse with a time gap of 14 or 15 days. Eclipses would help them to determine the length of the solar year and develop calendars based upon it.

Through assiduous record keeping passed from one generation to the next, after a few saronic cycles the patterns would be noticed by skilled astronomer-mathematicians. Even the fact that about one-third of all lunar eclipses are missed because the Moon was not visible at the appropriate time would become obvious.

It would thereafter become apparent that both lunar and solar eclipses occur only in allowed periods lasting for several weeks, and spaced by about 173 days (half an eclipse year). Poring over eclipse records, fastidiously maintained for generations by their predecessors, the ancient scholars would notice that eclipses of the same basic characteristics seemed to recur with gaps of 18 solar years plus 10 or 11 days (close to 19 eclipse years). Such sequences might last for a millennium or more. If instead 19 solar years were taken as the yardstick, short sequences of eclipses of varying types would be identified as occurring on or about the same dates within the calendar.

The recognition of these patterns of eclipses in the archives then would have allowed them to reverse the arrow of time, and project the cycles into the future. Eclipses could be predicted with utmost precision, without the need for an understanding of celestial mechanics and the use of elaborate calculations on an electronic computer. A few sums scratched on a papyrus scroll would do the trick. And a very useful trick it was, too.

Knowledge of the characteristics of the eclipse cycles enabled ancient astronomers to predict repeat performances without understanding even that the Earth circles the Sun and the Moon orbits the Earth. (In just the same way one can predict when song-birds will trill again in some Appalachian forest after a winter migration to Mexico, even though their route and homing instinct is not known. Similarly we know when salmon will return to spawn in the rivers where they hatched in the Pacific Northwest, although we do not understand how they find their way from the deep oceans where they live the majority if their lives.) Such knowledge of eclipses was *power*, for the magi who understood the cycles, and for the kings and emperors who employed them. We will explore this power of prediction in the next chapter.

3

Making Predictions

As there was going to be an eclipse on his birthday, through fear of a disturbance, as there had been other prodigies, he put forth a public notice, not only that the obscuration would take place, and about the time and magnitude of it, but also the causes that produce such an event.

Dion Cassius, writing about the solar eclipse of A.D. 45,
which occurred on the birthday of the Emperor Claudius

I n the preceding chapter we saw how eclipses occur in repetitive cycles. One can easily calculate these cycles, given prior knowledge of the lengths of the various types of month, and the year. The ancients did not have that prior knowledge, though. They tackled the matter from the other end: we have precision measurements from which we can deduce the eclipse cycles, whereas they recognized the cycles from their long-term observations, and from them deduced the month and year lengths. This is the reverse process.

In a similar vein, nowadays astronomers who study celestial mechanics (that is, the movements of celestial objects such as planets, satellites, comets, asteroids, and stars) mostly employ sophisticated computer codes. However, the modern era in which great advances were made in the study of the motion of the Moon was the last few decades of the nineteenth century, when no computers were available. The theories for the Moon's orbit were largely *analytical*, rather than *numerical*; that is, they involved long strings of trigonometric functions that describe the various relationships between angles such as the celestial longitudes and latitudes of the Sun and Moon.

The best-developed lunar theory was that of British mathematician Ernest Brown, who worked much of his life at Yale University. It contained in all 1,500 separate terms; to ascertain theoretically the position of the Moon at some stated instant, the equations involved cover several pages.

One might wonder why this is the case. The answer is that precision requires many distinct effects to be accommodated. To begin with, the orbit of the Moon is not about the center of the Earth, but about the barycenter, which is the center of mass of the Earth–Moon system. (For more information on this, see the Appendix.) The barycenter moves around because the lunar orbit is not circular, and its shape alters cyclically.

Next one must take into account the numerous perturbations of the Moon imposed by the gravitational tugs of large masses other than the Earth. In fact the major perturbation, producing about 99.99 percent of the variation in the lunar orbit, is due to the large attraction of the Sun. But the remaining 0.01 percent is significant. Several distinct classes of perturbation contribute to this. These include the shapes of the Earth and the Moon (neither body has a uniform distribution of mass, producing gravitational anomalies) and the presence of the other planets in the Solar System (each perturbs the lunar orbit directly and also has an indirect effect through its tugs on the Earth). Obviously the complete analysis is very complicated.

Such investigations were conducted before Einstein published his General Theory of Relativity, which was a step forward from Newton's Theory of Gravitation. Incorporating relativistic effects, and ultra-precise measurements from laser ranging and other modern technology, the latest computer-based lunar ephemeris contains about 7,000 terms, although even that is a misleadingly small number because of such things as the planetary positions needing to be calculated separately. ("Ephemeris" is a word used to refer to tables of positions of heavenly bodies; it is derived from the Greek word for a day. If you want to know where to look for a comet in the sky tonight and tomorrow, you need its ephemeris. And things that do not last for long, like a mayfly, are said to be "ephemeral.")

Clearly, modern knowledge of the motion of the Moon is hugely complicated. Only a subset of this collection of data is required in order to foresee eclipse occurrences in a vague manner. To predict the path of totality of a solar eclipse to within a fraction of a mile on the ground, however, necessitates a very complete understanding of how the Moon moves relative to the Earth and the Sun.

Humankind has built up that understanding over the eons first and foremost by observing phenomena accurately, and then recording the observations assiduously. To pick up the migrating songbird analogy again, we are at a similar stage in developing our comprehension of how their homing instinct works as were the inhabitants of Mesopotamia 3,000 years ago in their burgeoning awareness of eclipses. The sport of homing-pigeon racing has been developing for over a century, and they have been used to carry messages for longer, but how the birds navigate is still beyond our ken. It may be something to do with the terrestrial magnetic field, but we need much more scientific information before we can claim to understand it completely.

Regarding eclipses, the long road to our present state of knowledge began, as we saw in Chapter 2, by recognizing that patterns exist, but the lengths of the cycles posed difficulties. Consider the Metonic cycle. One could quite quickly determine the length of the synodic month by counting the days between full moons. To get a reasonably accurate evaluation you might do that for 20 or 30 months and then take the average. But the length of the year is another problem. Yes, many things recur seasonally, like the flowers sprouting each spring, but even counting the days spanning a couple of dozen consecutive springs can lead to imprecise year lengths owing to the vagaries of the weather. One could chart the sunrise, and note the time between visits to its southernmost rising point at the winter solstice, but around the solstice this does not alter much from day to day. The Sun moves faster in terms of its rising point around the equinoxes, when in theory it rises due east. However, there is only one chance a day to mark where it rises, and it may jump over that specific horizon point in the east, meaning that your derived year length will be inaccurate on the scale of a fraction of a day.

Other ways to measure the year are manifold. The Egyptians had two. One was when the bright star Sirius appeared again in the predawn sky, having been lost in the solar glare for a couple of months. This is called its *heliacal rising*. It occurs around mid-July, hence the term "dog days" for the hottest days of summer (Sirius being known as the Dog Star). Around that time of year the great inundation of the Nile would start. This annual event allowed the Egyptians an alternative method to measure the year, although hardly very accurate unless averaged over many decades. Despite realizing the year to be about 365.25 days long, the Egyptians persisted in using a calendar with precisely 365 days every year. The result of this was that the dates of the heliacal rising of Sirius and the flooding of the Nile shifted through the months on a cycle that took 1,461 years to complete. This is called the *Sothic cycle*, Sothis being the Egyptian name for Sirius.

It was relatively easy for ancient civilizations to deduce that the solar year was 365-and-a-fraction days in duration, but to recognize the coincidence of the Metonic cycle (that 235 synodic months is very close to 19 solar years) required diligence. To discover the precession of the equinoxes (the backwards movement of the equinox on a cycle taking 25,800 years to complete, as detailed in the Appendix) required a much better knowledge of the length of the year. Better measurements were needed than those that might be derived merely from watching seasonally repeated phenomena like bird flights, floods, or flowers.

Over many centuries the Babylonians and other ancient civilizations recorded their eclipses. Unlike in the modern era, when daily newspapers, magazines, and other media publish all the minutiae of life, ancient annals tend to be brief and abrupt, recording only the most notable events. For instance, they might include such mentions as "In that year a bright comet was seen, King Aaron died and was succeeded by his son Beta, and an earthquake caused great damage in the city of Mammon"; or "In the following year the Emperor Xenophon defeated the rebel leader Yahoo in battle near the river Zingiber; three months later a great eclipse of the Sun was witnessed throughout the land." It was such eclipse records that provided the requisite framework for the year to be determined.

THE JACQUARD LOOM

Until a couple of decades ago, computer programs were generally punched onto 80-byte cards, the cards dating back to Herman Hollerith, who introduced a machine in the late nineteenth century to process the information resulting from a population census of the United States.

The basic idea of coded cards came earlier. Nowadays, placards displayed in the windows of haberdashery shops may advertise multicolored beach towels or the like as having a "Jacquard weave." That is, the pattern is not merely printed onto the material; rather it is woven into the fabric. It was a Frenchman, Joseph-Marie Jacquard (1752–1834), who invented the first loom capable of producing such designs.

But how did the Jacquard loom manipulate the weave? That is, how did it instruct which longitudinal threads to move upwards, and which down, as the bobbin carrying the cross-thread in the weave shuttled from side to side? The answer is that the instructions were carried by a series of holes cut into flat tablets of wood, a hole in a specific position causing a particular thread to be raised, whereas unpunctured wood had the effect of making the thread drop.

An equivalent system is the punched-hole stack of connected cards used in a pianola, or the rotating slotted-metal disk in a nickelodeon, where the music is being played in response to the arrangement of the holes. Similar principles are at work in many fairground organs and the like.

There is a specific link to the development of computers here. If Charles Babbage had ever managed to complete the "analytical engine" that he began in the 1830s, it would have been the first programmable computer, although a mechanical rather than electronic device. Babbage, an Englishman, disparaged his own country greatly, but was a great admirer of what he saw as the superior ingenuity of other Europeans. He knew all about Jacquard looms. Babbage's intention was to read both data and program instructions into his machine using a card system copied from the Jacquard concept.

This is connected with eclipses in two ways. The first is that Babbage's specific initial motivation was the automated computation of mathematical and astronomical tables, such as might be used to predict eclipses. His initial fledgling device, begun a decade or so earlier, which again was never completed, was the "difference engine," a straightforward calculating machine rather than a programmable computer. Its development was funded in part by the British government on the grounds that the nautical almanac used for navigational purposes by the Royal Navy and merchant shipping was rife with anomalies. These were due to mistakes made in the complicated calculations performed longhand by human computers, rather than the error-free machines that Babbage claimed he would be able to construct.

The second point connecting to eclipses is that a Jacquard weave provides an excellent parallel to the patterns of eclipse occurrence.

ECLIPSE CYCLES AS WOVEN PATTERNS

Imagine the eclipse records from many centuries as being analogous to a vast woven pattern hung out on a wall, a tapestry of great complexity. One could visualize a color-coding of the threads for different types of eclipse: gold for total solar eclipses, silver for lunar, ruby for annular eclipses, sapphire for partial, in all manner of tones and hues.

Whole sections of records, equivalent to decades of time, may be missing due to miscreant scribes, fires in libraries, or national upheavals leading to disruptions in official diary keeping; these are like sections of cloth missing. Many eclipses will not have been seen due to geographical considerations, but that is like having moth-eaten holes in the cloth, with small parts of the pattern having been deleted. Similarly, some eclipses may be wrongly dated in some way, because of mistakes in copying annals; this is analogous to ink or dye accidentally spilled onto the tapestry, adding spots where none should be. However, the overall repeating pattern, the big picture, will still be clear.

Visualize this imaginary tapestry filling a wall facing you, right up to the corner, and then bending around it out of sight. The corner itself can be taken to equate to the present time, with the tapestry facing us representing the past, the section around the corner representing the future. The pattern we can see is beautiful, but repetitive, the same complicated cycles recurring, and so we know what lies around the corner, just as when we pull cloth off of a spool we can predict how the pattern will appear. Similarly, without having detailed knowledge of celestial mechanics, or computers following orbits with utmost precision, we can predict when eclipses are due to take place.

To provide an example of the sort of pattern that results, in Figure 3-1 all the solar eclipses that have taken place, or are due to take place, between 1901 and 2100 are plotted. Similarly all the lunar eclipses (neglecting the penumbral events) during that period are depicted in Figure 3-2. Those are our eclipse tapestries.

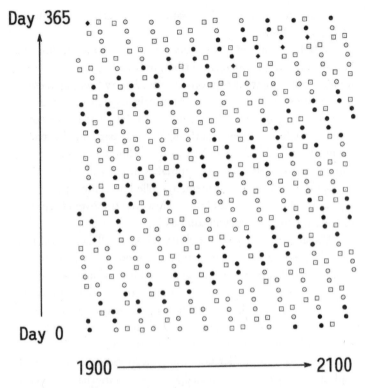

Day 365

Day 0

1900 ————————→ 2100

FIGURE 3-1. The pattern of solar eclipses between 1901 and 2100 is shown here. Diligent eclipse record keeping might have allowed an ancient society to predict future events, although this is more straightforward for lunar eclipses, as in Figure 3-2. (Solid circles represent total eclipses, open circles annular eclipses, and black diamonds hybrid events—part annular/part total. Partial eclipses are shown as open squares.)

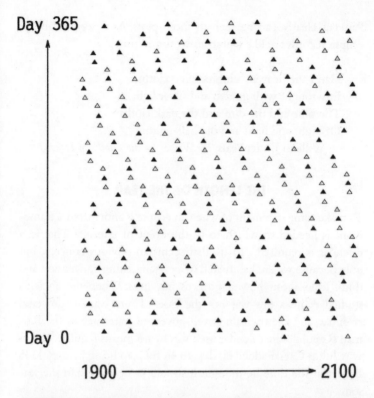

Day 365

Day 0

1900 ——————————▶ 2100

FIGURE 3-2. The pattern of lunar eclipses between 1901 and 2100 plotted in a similar way to Figure 3-1. These data are based upon our modern knowledge of the lunar orbit. However, in principle something like 60 percent of all the lunar eclipses over an extended period might be charted by some ancient civilization simply from diligent sky-watching, assuming that cloud cover sufficient to blanket the Moon for some hours did not occur throughout the kingdom. (Solid triangles represent total eclipses; open triangles are partial events.)

Patterns clearly can be seen in those plots. As described above, eclipse records are like tapestries woven in time:

> Had I the heavens' embroidered cloths,
> Enwrought with golden and silver light,
> The blue and the dim and the dark cloths
> Of night and light and the half-light...
>   William Butler Yeats, *He Wishes for the Cloths of Heaven*

THE LENGTH OF THE YEAR

To make some statement of when an eclipse is anticipated, a framework is needed to which to fit the event of interest. That is, a calendar is required. To us, that seems an obvious concept, but only because we are habituated to a certain dating convention and think little about how it governs our lives. When the ancients studied eclipses this was not the case. There was no universal calendar, and even within a well-governed state such as the Roman Republic the calendar used was by no means regular. That is why Julius Caesar added 80 days to 46 B.C., to bring January 1, 45 B.C. near the time he thought it should be according to the seasons.

The situation was similar elsewhere. Although King Ptolemy III of Egypt had decreed in 238 B.C. that a quadrennial cycle of leap years should be used, to reflect the real duration of a year, his dictate was not put into common use. Different nations used calendars that drifted against the seasons, such drift either being allowed to continue, as with the earlier Egyptian Sothic cycle, or being abruptly corrected from time to time, as with the Roman calendar before Julius Caesar rectified matters.

The design of any perennial calendar obviously would require a detailed knowledge of the year length far beyond the flowers and floods mentioned earlier. The tapestry laid down by the eclipse records made this possible. Imagine that one vertical thread in our wall-hung tapestry represents a year and that you have somehow managed to get the year length correct. Time starts at far left and proceeds to the right until the present, which is where the tapestry turns a convex corner and is not yet visible to us.

Under this circumstance of one thread per year we can begin to pick up some features of the pattern that must result. The 19-year Metonic cycle produces numerous sets of four or five dots arranged horizontally. Each individual dot will be a component part of one of the sequences of 70 or 80 dots produced by the saronic cycle, these being slanted from bottom left towards upper right because their time spacing is 18.03 years. The origins of various other components of the tapestry pattern are detailed in the Appendix. For example, there will be sharply downwards-sloping lines produced by eclipses coming 10.88 days earlier from year to year, because 12 lunar months are that amount short of a full solar year.

It is not necessary to continue with more features of the pattern, such as the 3.8-year gaps (a subdivider of the 19-year Metonic cycle). You can refer to Figures 3-1 and 3-2 to see what I mean. The thing to recognize here is that if an incorrect length for the year were employed, then the complex pattern of the tapestry would be skewed. Getting the tapestry straight and agreeable provides in effect a precise evaluation of the length of the year. This measure is far more accurate than floods or flowers, cuckoo calls or salmon spawning, sunrises at solstices or equinoxes, or waiting for Sirius to emerge again from the glare of the Sun, as we shall see below.

THE ECLIPSE GRID

Records of eclipses provided a "grid" against which the lengths of the celestial cycles could be reckoned, because eclipses repeat on a variety of distinct cycles defined by the lengths of month and year. You may be familiar with simple regularly spaced grids, like graph paper divided into big squares one inch on a side, and smaller squares each one-tenth that. The eclipse grid is different. The great thing about eclipses is that the subdivisions are not so simple, and that allows greater precision.

Similarly in many trades like engineering some device such as a vernier caliper is used. A vernier has sliding scales on which divisions are marked unequally on the two sides: on one side a centimeter may be split into ten separate millimeter marks, whereas on the sliding part which faces it the centimeter is divided into just nine equal units (each of 1.11111... millimeters). By noting where two notches align, it is possible to measure lengths accurate to one-hundredth of a centimeter, a tenth of the smallest division.

The eclipse grid provides not just one set of overlapping measuring sticks, like a vernier, but many. Because of that it can be used to deduce not only the duration of the year, but also the lengths of the various types of month, even without having artificial clocks available. The ancient Babylonians, Greeks, and Chinese, remember, did not have the advantage of mechanical timepieces, only water clocks and sundials.

Put yourself in their place. Imagine that you note a partial solar eclipse beginning about six hours after midday, a time that is determined simply by measuring the angle of the Sun from the noon meridian. Looking back through the records you find that a similar event occurred 18 years and 10 days before, at about two hours before noon. There have been 223 synodic months in between, and you can determine the relative time of day accurate to about half an hour. This enables you to stipulate the mean length of a synodic month to better than a minute. This is much, much better than one could achieve simply by trying to judge when the Moon is fullest, especially as no clock is available. The other types of lunar month, discussed in the Appendix, may similarly be calculated.

That is how the month duration could be reckoned using eclipses. Now what about the year? The precession of the equinoxes was discovered by Hipparchus, a Greek astronomer who lived between about 190 and 125 B.C. The task Hipparchus began with was a determination of the lengths of the year and the months.

Others had come before him. In 432 B.C. Meton had proposed his 19-year cycle containing 6,940 days, producing an average year length of 365.2632 days. About a century later his countryman Callippus advocated an alternative cycle four times as long. His cycle of 76 years he thought should contain 27,759 days. That is one less than four Metonic cycles (27,760 days) and is a better approximation to the real year length: 27,759 divided by 76 equals 365.25. It is not perfect, though. When Hipparchus made his determination of the year another two centuries later, he arrived at a value one three-hundredth of a day less. Noting that four Callippic cycles last for 304 years, Hipparchus proposed a cycle of that duration with one day subtracted to compensate for the year really being slightly shorter than 365.25 days.

Actually, the year length derived by Hipparchus was still slightly wrong. (One part in 300 of a day below 365.25 results in 365.2467 days, whereas the mean tropical year—the average year length used by astronomers, as discussed in the Appendix—of 365.2422 days is about one part in 128 below.) Nevertheless his was a remarkable achievement because he was able to show that the seasonal cycle was not the same as the time between the stars returning to the same places in the sky, called the *sidereal year*. That is, Hipparchus recognized the movement of the Earth's spin axis known as the precession of the equinoxes. In view of that his name is revered in the history of astronomy.

From our present perspective we should ask how he did this. The answer to that is simple: through eclipse tables. Hipparchus made his own eclipse observations between 146 and 135 B.C., and compared these with earlier Babylonian records. *That* is how he was able to determine the years and the months so accurately, feeding into the knowledge base that eventually brought about our calendrical systems.

Hipparchus lived well before the invention of the mechanical (let alone electronic or atomic) clock, the telescope, or finely divided measuring scales, pocket calculators, and computers. He was working two millennia before the necessary physical theories were developed allowing the motion of the Moon across the sky to be programmed and thus calculated ahead of time with utmost precision. It may well be that unknown Babylonian astronomers had beaten Hipparchus to it, using the same techniques, but that does not diminish the stature of his work. In ancient times, then, various individuals of genius, living in societies possessing careful records of past celestial events, were able to interpret those records and deduce the lengths of the years and the months to a matter of minutes and seconds. That was a considerable achievement, made possible only by the regularity of the eclipse grid.

BY THE RIVERS OF BABYLON

Much of the eclipse and calendar knowledge spreading from the Middle East to Greece and Rome and thence the rest of Europe stemmed from understandings developed in Mesopotamia between 3000 and 500 B.C.

Mesopotamia is strictly the region between the Tigris and Euphrates rivers. In ancient times it was a bountiful shallow valley, the home of several distinct civilizations over the last three millennia B.C. Babylon itself was on the Euphrates, about 60 miles south of modern-day Baghdad. The establishment of the city predates 3000 B.C., and by 2500 B.C. the entire region was united under Babylonian rule.

The early peoples of that region are generally termed the Babylonians, but one should be aware that there were racial and cultural differences as power changed hands from one era to the next. Much of the learning of the melded culture that arose came from the Chaldeans, who originated on the western side of the Euphrates. Near where that river formerly emptied into the Persian Gulf was the city of Ur, the capital of the Sumerians, who lived along the northern fringe of that sea. From east of the Tigris came the Elamites, and from the north of Babylon arose the Akkadians. All these may be subsumed into the overall Babylonian Empire, the heights of which were reached between 2800 and 1700 B.C.

In the following thousand years their power ebbed, while the bellicose Assyrians from farther north became the dominant culture, conquering Babylonia in 689 B.C. and destroying much of the city. Thankfully the Assyrians did not obliterate the long-standing astronomical culture of Babylonia. They soon adopted various superstitious practices based upon the belief that celestial phenomena were harbingers of approaching events on Earth. The major developments that led to horoscopic astrology occurred in this era. Comets and vivid shooting stars were interpreted variously as being auspicious or dangerous omens, while eclipses were regarded as being highly significant. An example is the following prophecy from a court astrologer: "On the 14th an eclipse will take place; it is evil for Elam and Amurru, lucky for the king, my lord; let the king, my lord, rest happy. It will be seen without Venus. To the king, my lord, I say: there will be an eclipse. From Irasshi-ilu, the king's servant." Obviously the ruler could be put in a good temper by having an eclipse interpreted in advance as being beneficial (but heaven help the astrologer should, say, the king's favorite horse or dog fall sick that day). Such predictions could be made with an incomplete understanding of eclipse cycles. The astrologers might notice sequences of several lunar eclipses recorded six full moons apart, and once the first in a new series was seen the subsequent events might be calculated. This is a much simpler knowledge than that of the saronic and longer-term cycles. The problem for the astrologers was that they could not anticipate the first eclipse in a series, and that might incur regal displeasure.

Assyrian rule was only temporary. Weakened by various incursions around its periphery, the over-stretched Assyrian Empire succumbed by 606 B.C. to attacks from the resurgent Babylonians and the Medes (the kingdom of Media was to the northeast, towards the Caspian Sea, the northwestern part of modern Iran).

Under the famous king Nebuchadnezzar, who ruled from 604

until 561 B.C., the Babylonian Empire expanded. They rampaged to the west and destroyed the great Temple of Solomon in Jerusalem, leading to the Exile (the captivity of the Jews in Babylonia, 597–538 B.C.). In this climate of astrological belief the Babylonian priesthood who read the signs of the sky became rich and powerful, the regents and generals making decisions based on advice interpreted from celestial phenomena, both of the past and anticipated in the future.

The Babylonian regime was overthrown for the last time, by the Persians, early in the fourth century B.C. When the Jews eventually returned to Judea, they took with them the astronomical knowledge on which the Hebrew calendar is based, with its strict rules for phasing various religious feasts against the Sun and the Moon. They had no time for the astrological deities of the Babylonians, but they did want to know about how the planets moved in the sky. In those days the term planets encompassed all regular moving objects: the Sun and the Moon, as well as Mercury, Venus, Mars, Jupiter, and Saturn. That makes seven. Our seven-day week derives from the astrological planetary week of the era reinforced by the Jewish Sabbath cycle of seven days.

With their new Persian masters the astrological priesthood in Babylon needed to adapt to preserve their privileged place in society, and to do that they needed to develop a better understanding of how the celestial objects moved. Studies of past eclipse records intensified, and it seems likely that about this time the saros was discovered.

There is direct evidence of this discovery. A fragment of an eclipse list between 373 and 277 B.C. has survived, and it is split into columns covering 223 synodic months; this is the number in a saros. A saros, remember, contains 19 eclipse years, each containing two eclipse seasons, making 38 in all. Each of the columns mentioned consists of 38 horizontal lines. It seems that the Babylonian astronomers knew about the saros at least by the third century B.C., and so were able to predict eclipses into the distant future rather than merely short-term runs.

By then the Persian Empire had been overwhelmed by Alexander the Great, and from about 331 B.C. Babylonia was incorporated into the vast empire that had been conjoined through his conquering forays west through Egypt, and then east all the way to India.

Alexander was from Macedonia, the northern part of what we now call Greece, as opposed to Athens and the southern states. His dynasty ruled much of the eastern Mediterranean for some centuries, for example as the Ptolemies in Egypt. The last of them was Cleopatra. After Alexander's death—in Babylon in 323 B.C., at age 33—the lands he had conquered were consolidated into what became known as the Seleucid Empire.

Under Greek hegemony Babylonian astronomy continued to thrive, and the results of observations were relayed back to Greece, to men such as Hipparchus. It was the Babylonian records of eclipses, coupled with his own observations, that enabled Hipparchus to take such major steps forward in determining the cycles of the heavens.

PROJECTING THE PATTERN FORWARDS

How far back do the eclipse records of Babylon go? Solar eclipse notations that may be unambiguously interpreted and dated start from 700 B.C., but most postdate 350 B.C. On that basis, assuming that at least a century of records would be needed to decipher the saros, it would seem unlikely that eclipse prediction based on those records would have been possible much before 250 B.C.

Who, then, was first to predict a total solar eclipse correctly? This is a question over which historians of astronomy have argued a great deal, because there is an apparent prediction from much earlier than that.

Herodotus (484–425 B.C.) was a Greek historian who wrote most of the surviving accounts of his era and earlier. He claimed that Thales of Miletus (see Figure 3-3) predicted the solar eclipse in 585 B.C. that occurred during a battle between the Medes and the Lydians. (Lydia was the western end of Asia Minor, where the city of Miletus was located.)

Thales does seem to have understood the rudiments of solar eclipses, recognizing that they are due to the Moon passing in front of the Sun, although in his day the nature of orbits was unsuspected. Thales thought of the Earth as a flat disk floating on a great sea, the Sun and Moon being other disks moving above it, and sometimes they happened to align. The suggestion of the Earth circuiting the Sun remained some time off. Aristarchus of Samos proposed the concept in the third century B.C., but it was not until after the Copernican revolution in the sixteenth century that the idea gained wider acceptance, in the face of ecclesiastical opposition.

The 585 B.C. eclipse certainly seems to have caused the Medes and the Lydians to reconsider their hostile intent and agree to a peace treaty after five years of war, each seeing it as an omen; however, it is not clear that Thales predicted its date and circumstances. We are able to back-calculate to show that the path of totality on the afternoon of May 28 swept along the Mediterranean and fairly centrally from west to east across Asia Minor, where

FIGURE 3-3. The pioneering Greek mathematician Thales is often credited with making the first prediction of a total solar eclipse, although historians of science now doubt whether he did more than suggest that such an eclipse would occur in a certain year. Thales was from Miletus, a town on the western coast of what is now Turkey. The eclipse track passed across that area in 585 B.C., bringing to an end a long-standing war between rival peoples.

the armed dispute was taking place and the Sun was blanked out for over six minutes.

It was a very unusual event, but Herodotus wrote only that Thales gave the year, so one might wonder whether it was a true prediction or just a lucky guess. Predicting that a partial solar eclipse will occur is fine, but getting a total solar eclipse right is another thing entirely. On balance it seems that Thales and his contemporaries did not know how to foresee eclipses by any means other than the short-term relations like the ten-day shifts from one year to the next. Hipparchus used eclipse data, and the saronic cycle, to ascertain accurate values for the year and the lunar months, but did not make forward eclipse predictions.

The eclipse knowledge gathered by the Babylonians lay dormant for many centuries. Hipparchus and others knew that the year was not exactly 365.25 days long, and yet the Julian calendar leap-year cycle based upon that length persisted until the sixteenth century. In the same way, the detailed cycles making eclipse prediction possible were not to be used for a long time.

EDMOND HALLEY AND ECLIPSES

The first real predictor of eclipses will come as a bit of a surprise. Edmond Halley (1656–1742) knew that the comet bearing his name would come back in 1758, long after his death, and said he hoped that when it did appear it would be recalled that it was an Englishman who had foreseen its return. But Halley has another claim to fame with respect to predictions: in the modern era it was he who recognized how to use the saros to pre-calculate eclipses. In fact, his contemporaries considered that he had discovered that cycle, not realizing that the Babylonians and Greeks had known of it so long before, the understanding having been lost. It was Halley who gave the saros its name.

From the late seventeenth century Halley was one of the lions of the Royal Society of London (see Figure 3-4). His scientific interests were many and various. In 1693 alone Halley read papers at meetings of the Society covering such disparate subjects as:

- How to determine the positions of the tropics
- The pressure within a diving bell
- How the length of the shortest day varies with latitude
- How deformed fingers are inherited within some families
- Mortality rates and annuities
- How crabs and lobsters regrow amputated claws
- A hydrographic survey of the coast of Sussex, in the south of England

FIGURE 3-4. Edmond Halley pictured in his younger days, shortly after he discovered the saros. The inscription shows that, apart from being a doctor of laws, Halley was also Savillian Professor of Geometry at Oxford University and Secretary of the Royal Society of London. Later he took up the appointment of Astronomer Royal.

Obviously he was a very busy man.

Halley's investigations of eclipses was a recurring theme, and the previous November he is recorded to have given ". . . an account of the Eclipses of the Sun and Moon to bee computed by an easy calculus, from the Consideration of the Period of 223 Months, shewing how to aequate between the extreams of the excess of the odd hours above even days, which is always between 6.20 and 8.50. He produced a Table ready calculated for this purpose, and shewed the use thereof. Which he promised to exemplify against the next Meeting." That, in effect, is the announcement of the discovery of the saros, Halley having recognized even the limits to the odd hours and minutes above any particular 18-year plus 10- or 11-day period. The following week "Halley shewed a Paper wherein he had computed the Eclipses of the Moon in severall Series, and said, that he found, that he could very well represent them all; much nearer than they were observed by the severall observers." How could one predict something more accurately than it could be observed? The answer is that Halley had found that lunar eclipses predicted using the saros provided a more precise timepiece than the mechanical clocks used by the observers, and for matters of navigation that was potentially a most valuable discovery.

THE NAVIGATIONAL UTILITY OF LUNAR ECLIPSES

Britain's rule of the waves from Halley's time onwards came about not only from its strong navy, but also through its scientists providing accurate navigational charts and methods for determining position at sea.

This did not happen overnight. The measurement of one's geographical longitude was a long-term problem. Deduction of the latitude was relatively easy, from the minimum angle achieved each day between the Sun and the overhead point (called the *zenith*). This minimum occurs at noon. At night, various stars can be used. Tables of Sun and stars were available allowing a ship's latitude to be ascertained in that way, but longitude is a different story.

As you sail east or west the time according to the position of the Sun alters. If you had an accurate clock that maintained the time at some reference point, say back in London, then by comparing the clock time with the time according to the Sun in the sky, the longitude might be determined. Unfortunately the pendulum clocks used in churches and observatories would not work on a tossing and rolling ship at sea.

In 1714 the British government offered a very large prize—

£20,000, worth about $3 million today—to anyone who could solve this general problem and enable ships to be navigated more safely. Prospective solutions fell into two camps. One approach involved constructing mechanical clocks that would function accurately on board ship, and this led to many advances in timekeeping. (The identity of the word for a time period spent maintaining a lookout, and a small timepiece that will fit in a pocket or strap to your wrist, did not come about by accident. I refer, of course, to a *watch*.) The problem was eventually solved this way by a skilled artisan, John Harrison, although there was much wrangling over the award of the prize (he never received the cash and credit which was his due) continuing for several decades.

Harrison was an outsider to the scientific establishment, which favored a different method: using astronomical objects as natural clocks. In principle, for instance, the positions of the four giant moons of Jupiter might be read as the hands on a clock, showing the same time whether viewed from anchor in the Thames estuary or from Port Royal in Jamaica.

Jupiter, though, could not be seen for much of the year when lost in the solar glare and was also difficult to observe telescopically from a ship in the mid-Atlantic. The Moon provided a better target. It could be seen at some stage during the day for all except about 72 hours straddling conjunction each month, and in principle its position could be used to give the time.

The problem was that the location of the Moon in the sky, from a theoretical basis, was not known with sufficient precision. The best available set of positions for the Moon computed in advance was derived from the lunar theory published by Sir Isaac Newton in 1702, but observations showed these to be inaccurate. Halley examined this question and, realizing that the eclipse grid allowed a major refinement, he suggested a solution that effectively used the saros.

Some decades before, John Flamsteed (1646-1719) had been Astronomer Royal and had made measurements of the lunar positions, these showing varying discrepancies from the positions according to Newton's theory. Between 1722 and 1740 (a complete saros) Halley, by then Astronomer Royal himself, made 2,200 observations of the same parameter, and discovered that the discrepancies charted against the theoretical positions simply repeated those displayed by Flamsteed's measurements from 18 and 36 years earlier. This indicated that Newton's theory could be numerically corrected using the saros in quite a simple way.

In the middle of his observations, in 1731, Halley recognized the potential of this method to provide a solution to the navigation problem, but failed to publish the results during his lifetime. By the time Halley's analysis appeared in 1749, better lunar theories had been developed. Unbiased observers also had realized by then the accurate and practical use of Harrison's clocks. This did not, though, stop the establishment astronomers from fighting a continuing rearguard action.

Edmond Halley's lunar observations were never used in the practical matter of navigation, but his earlier investigations did lead to the rediscovery and naming of the saros. Halley recognized not only that eclipses repeat on that cycle, but also that the eclipse characteristics recur. To that extent he is the true father of eclipse prediction as we have received it.

EARLIER USAGE OF SHORT ECLIPSE SEQUENCES

Although the saros had been forgotten between the era of the Babylonians and Greeks and Halley's time, the fact that short-term sequences of eclipses occur had not. Perhaps it might be more correct to say that each age rediscovered such coincidences, just as generations of schoolchildren look at their atlases, note that South America could be shifted eastwards and twisted to fit rather nicely into the concavity of Africa, and thus reinvent the concept of "continental drift." Regular sky watchers would soon realize that eclipses tend to repeat in series moving progressively earlier by ten or so days in the year, such that the next event might be predicted. Similarly the Metonic cycle was well known, providing a 19-year pattern, plus the 3.8-year subdivider.

Some forward-prediction of eclipses over decades was feasible in medieval times, then, although it awaited Edmond Halley to tease out the secrets of the saros, employing the gravitational theory of Newton plus other achievements of the burgeoning pursuit of natural science. Three centuries before Halley and Newton, an astronomer might gather eclipse records from manuscripts kept in monasteries and identify patterns, but the wide dissemination of eclipse predictions could not occur until the introduction of printing.

Johannes Gutenberg (1400–1468) is normally credited with the invention of the printing press. It was another Johannes, also a German, who in 1472 became the first person to print an astronomical almanac. This was Johannes Müller, better known as Regiomontanus, the latinized name of the city of Königsberg where he was born. Regiomontanus produced printed predictions of when eclipses were due, and these tables plus later works of a similar nature would prove to be important for navigational purposes. Consider an example.

In the 1580s the British wanted to found a new colony in North America, that colony eventually becoming Virginia, named for Queen Elizabeth I who sanctioned Sir Walter Raleigh's tentative exploration of the region. The first thing they needed to do was to determine the geographical coordinates of the area, so that later ships would be able to find their way. Basically, Raleigh and his colleagues needed to know the width of the Atlantic Ocean.

It was known from the tables that a total lunar eclipse was due at about midnight (London time) on November 17–18, 1584. And so a pair of astronomers and their assistants landed on Roanoke Island, just off the main coast (where they might be protected to some extent from hostile natives), some months ahead of time. The eclipse would be visible both from England and the west of the Atlantic.

By setting up a pendulum clock and synchronizing it with the local time according to the Sun, the astronomers were able to say when the eclipse started as they saw it. At precisely the same instant astronomers in England would note the onset of the eclipse according to their own clocks. The difference in the times reflects the difference in longitudes, and thus the coordinates of the island were calculated once the data were brought back to England. Knowing that one location, it was then simple to determine other points in the new colony, in the same way as we might refer directions to some local landmark (like "five blocks west of Grand Central Station").

An important factor to note is that only lunar eclipses were of utility in this regard. A lunar eclipse could be seen from the entire night-side hemisphere, the instants at which the various contact points are observed being independent of the viewer's location. This is not the case for solar eclipses: the contact times in that case depend critically upon your location, the Moon's shadow taking some hours to sweep across the globe.

The obvious usefulness of lunar eclipses for ascertaining the longitudes of transoceanic reference points meant that most voyagers carried predictive tables of such events. A prime example is Christopher Columbus, who possessed a copy of the *Calendarium* published by Regiomontanus in 1474. Most people know that Columbus landed in the New World in 1492, but few realize that he made several subsequent transatlantic trips. An eclipse saved him and his men on the fourth of his westerly ventures.

CHRISTOPHER COLUMBUS AND THE LUNAR ECLIPSE OF 1504

Columbus struck trouble in the Caribbean in 1503 when, having already needed to abandon two ships, his last pair of caravels also became riddled with marine worms. He was forced to lie up on the northern shore of Jamaica, at a small cove named Santa Gloria (now Saint Ann's Bay).

The Jamaican indigenes were friendly when Columbus arrived, but their hospitality had begun to wane after six months of the prolonged Spanish stay, the stranded party repeatedly needing to request food and water in return for such trinkets as they could offer, things like beads, nails, and mirrors. Both the novelty and the supply had run out by the end of the year.

The admiral had sent a party of men east in small boats to the Spanish-occupied island of Hispaniola (now Haiti and the Dominican Republic) to seek help, but did not hear back. In January 1504 half of the remaining crew mutinied and departed for Hispaniola, attempting to make the hundred-mile passage in canoes hewn from local timber.

This left Columbus with 50-odd men on board two worm-permeated vessels. He could not abandon the ships because of the many valuable items on board, not the least being the survey maps he had drawn up in exploring the coasts of Honduras, Costa Rica, Panama and Nicaragua as he searched unsuccessfully for a passage west to the Pacific and Asia. By February the Indian caciques (leaders or chieftains) saw the Spaniards were at their mercy and refused to provide any more provisions.

Columbus was desperate. Referring to his *Calendarium* he found that a total lunar eclipse was due on the evening of February 29 (soon after midnight on March 1 as seen from Europe). He invited the caciques on board his flagship, the *Capitana*, providing them with some entertainment but with serious undertones. Columbus explained that he and his men were Christians who worshipped a powerful god, superior to the deities of the Jamaicans, and that He had been angered by their refusal to succor the Spaniards in their time of need. As a result it was the intention of God to punish them with famine and disease, but He would give the caciques one last chance, by providing a sign from Heaven of His displeasure, darkening the full moon soon after it rose in the east. As an additional clear indication of divine wrath, the Moon would be reddened. If they paid heed and changed their ways they might be saved from pestilence and starvation. With this Columbus sent them on their way.

Many of the chiefs mocked Columbus for his suggestion, but others were less confident. As the Moon climbed above the horizon it was seen to be somewhat dimmed, the partial eclipse having already begun. All were convinced as the shadow of the Earth enveloped the orb rising in the east, reaching totality an hour after moonrise. Pandemonium ruled, and the caciques dropped to their knees, begging Columbus to intercede on their behalf and save them, as depicted (rather imaginatively) in Figure 3-5.

Columbus was too smart to agree immediately. For added effect he retired to his cabin, knowing that the total phase would last for about one and three-quarter hours. Having timed his withdrawal with a sandglass, Columbus reemerged at the appropriate time. He told the Jamaicans that he had consulted God and persuaded Him to cease the shielding of the Moon, so long as they promised to behave themselves and supply the Spanish for so long as they needed to stay. The caciques hastily agreed, and with a wave of his arm Columbus gave the sign that the Moon should be unveiled, which of course was promptly enacted in the sky as the shadow slowly receded.

The Spaniards still needed to wait until June before a rescue ship appeared, but they did not lack food or water during the interim. For Columbus the eclipse had another implication, because it made it possible to calculate his longitude.

FOOLING THE NATIVES?

This tale of Columbus's deceptive use of an eclipse to fool a less scientific people has been echoed in various works of fiction. Quite likely the episode provided a direct inspiration for such writers; for example, Washington Irving recounted Columbus's subterfuge in a best-selling book, making the story well-known.

In 1889 Mark Twain published *A Connecticut Yankee in King Arthur's Court*, a novel that envisions life in sixth-century England. The author has Hank Morgan, the Yankee in the title (and Bing Crosby in one movie version), hoodwinking the ignorant folk of that era by invoking prior knowledge of a solar eclipse due on

FIGURE 3-5. Christopher Columbus is begged for forgiveness as he invokes the power of the Christian God to eclipse the Moon, persuading the Jamaican natives that it would be wise to supply his party with food and other necessities.

June 21, 528, even stating the precise time of totality (three minutes past noon). Twain has Morgan, who is jailed awaiting execution, threaten King Arthur with a blanking out of the Sun:

> Go back and tell the king that at that hour I will smother the whole world in the dead blackness of midnight; I will blot out the Sun, and he shall never shine again; the fruits of the Earth shall rot for lack of light and warmth, and the peoples of the Earth shall famish and die, to the last man!

Morgan, though, is not believed, and he is tied to a stake to be burnt, Merlin wanting to light the flames himself. As in any thriller, rescue comes in the nick of time:

> I said to myself that my eclipse would be sure to save me, and make me the greatest man in the kingdom besides . . .
>
> I waited two or three moments: then looked up; he was standing there petrified. With a common impulse the multitude rose slowly up and stared into the sky. I followed their eyes; as sure as guns, there was my eclipse beginning! The life went boiling through my veins; I was a new man! The rim of black spread slowly into the Sun's disk, my heart beat higher and higher, and still the assemblage and the priest stared into the sky, motionless. I knew that this gaze would be turned upon me, next. When it was, I was ready. I was in one of the most grand attitudes I ever struck, with my arm stretched up pointing to the Sun. It was a noble effect . . .
>
> "Name any terms, reverend sir, even to the halving of my kingdom; but banish this calamity, spare the Sun!"
>
> My fortune was made. I would have taken him up in a minute, but I couldn't stop an eclipse; the thing was out of the question. So I asked time to consider. The king said:
>
> "How long—ah, how long, good sir? Be merciful; look, it groweth darker, moment by moment. Prithee how long?"
>
> "Not long. Half an hour—maybe an hour."

There were a thousand pathetic protests, but I couldn't shorten up any, for I couldn't remember how long a total eclipse lasts.

It grew darker and darker and blacker and blacker.... It got to be pitch dark, at last, and the multitude groaned with horror to feel the cold uncanny night breezes fan through the place and see the stars come out and twinkle in the sky. At last the eclipse was total, and I was very glad of it, but everybody else was in misery; which was quite natural.... Then I lifted up my hands—stood just so a moment—then I said, with the most awful solemnity:

"Let the enchantment dissolve and pass harmless away!"

There was no response, for a moment, in that deep darkness and that graveyard hush. But when the silver rim of the sun pushed itself out a moment or two later, the assemblage broke loose with a vast shout and came pouring down like a deluge to smother me with blessings and gratitude.

Twain's description of the eclipse seems accurate in every way, except one. There was no solar eclipse visible in England in A.D. 528. That was an invention.

One must never let the facts get in the way of a good story. In his first novel, *King Solomon's Mines* (1886), H. Rider Haggard has his heroes escape the clutches of a despotic African king by using a predicted eclipse in a similar way. Mind you, Haggard could not make up his mind whether it was a solar or a lunar eclipse, changing from one to another between editions: "Yet I tell you that tomorrow night, about two hours before midnight, we will cause the Moon to be eaten up for a space of an hour and half an hour. Yes, deep darkness shall cover the Earth, and it shall be for a sign." The lunar eclipse duly occurred, and while the natives are in terror of their lives (Figure 3-6) Allan Quatermain and his colleagues make a getaway. In the previous edition it was a solar eclipse just after midday. Perhaps someone had told Haggard that his science was wrong; he has Quatermain describing their flight in this way:

For an hour or more we journeyed on, till at length the eclipse began to pass, and that edge of the Sun which had disappeared the first became again visible. In another five minutes there was sufficient light to see our whereabouts . . .

Many eclipse enthusiasts would love to suffer the slings and arrows of hour-long totality, but the laws of physics forbid it. A handful of minutes is all you can get.

This basic idea of using an eclipse to escape hostile but ignorant natives was copied by the Belgian writer Hergé (Georges Rémi) in his *Adventures of Tintin*. In one episode Tintin and his eccentric colleagues are to be burnt at the stake by the Emperor of the Incas, having tried to make off with their pockets full of diamonds, just as in *King Solomon's Mines*. Although his friends think that Tintin is babbling nonsense, in fact he is giving praise to the Sun as an eclipse approaches, bringing about their salvation.

MODERN TIMES

It is clear that eclipses have had an importance in the development of human society far beyond people simply wondering at their origin. Eclipses provided the measuring stick with which the year was determined, resulting in accurate calendars. From time to time they startled the ancients, perhaps precipitating pivotal moments in history, as the darkening of Sun or Moon was seized upon as a

FIGURE 3-6. In King Solomon's Mines the author H. Rider Haggard described his British heroes escaping from their African captors by using an eclipse listed in their almanac. Here the natives stare terrified at the darkened Moon, while the former captives make good their escape.

propitious omen by some wily commander or feared by a superstitious enemy. The ability to foresee when eclipses would be witnessed allowed more scientific cultures to impose their will upon others, as in the case of the subterfuge conducted by Christopher Columbus.

Seeing the curved profile of the Earth cast onto the Moon, the ancient Greeks reasoned that the planet is spherical, and this was backed up by other simple observations like the finite curved horizon espied from the top of a mountain. The eventual acceptance of that notion, and the Earth's movement about the Sun in common with the other planets and celestial wanderers like comets, led at last to an understanding of the lunar motion. From that comes our ability to predict eclipses independent of any past record: I can run a computer program using the lunar and solar orbits, printing out when the bodies align with utmost precision, without direct reference to any past eclipse.

Such a computer program may be complicated, but basically it uses the simple gravitational theory of Isaac Newton. We have studied the eclipses of the past, through to Newton's time, but now we might like to come a little more up to date. That involves an interstitial step, though, in which eclipse observations were employed to show that although Newton's theory is a good approximation, it is not a one hundred percent accurate description of the universe.

About the time Charlie Chaplin was making his earliest movies, eventually culminating in *Modern Times*, eclipse observations were likewise starting to enter their own modern times and being used to verify Albert Einstein's Theory of Relativity. That is the subject to which we now turn.

4

A Warp in Space

By means of prolonged processes of mathematics, entirely separate from the senses, astronomers are able to calculate when an eclipse will occur. They predict by pure reason that a black spot will pass across the Sun on a certain day. You go and look, and your sense of sight immediately tells you that their calculations are vindicated. So here you have the evidence of the senses reinforced by the entirely separate evidence of a vast independent process of mathematical reasoning.

Sir Winston Churchill

The Sun shines—but *how?* A hundred years ago this deceptively trivial question was causing great consternation not only to astronomers, but also to other scientists.

During the previous centuries, Western scientists discovered new phenomena that raised previously unsuspected quandaries. For example, before we realized that biological evolution occurs, producing new species from old, the avenue by which genetic change takes place was not a problem for consideration. Today, five or six generations after Charles Darwin, the mechanisms and processes of natural selection remain hotly debated within the academic community.

The question of how the Sun shines—that is, the source of its energy—did not become a matter of concern among scientists until the concept of geological deep time was established. Many will have heard of Archbishop James Ussher and his seemingly absurd statement that the world began in 4004 B.C. Those who mock Ussher do so from their own ignorance. One should not judge him by the standards of modern-day scientific knowledge, but rather from the perspective of the accepted wisdom in his own time, the mid-seventeenth century. In those days the age of the Earth was thought to number only a handful of millennia, and Ussher's conclusion was a respectable effort in the context of the scholarship of his era.

The realization that our planet is not just millions, but actually *billions* of years old was a long time coming. Edmond Halley enters our story again at this juncture: he suggested that the age of the Earth might be estimated by comparing the salinity of rivers with the salt content of the oceans, reasoning that the saltiness had built up over the eons. There are various shortfalls with this concept, but later experimenters did derive ages of many millions of years based on such measurements.

Another method was founded upon the observation that far below ground, deep down mine shafts, the rock is hotter than at the surface. Volcanoes provide unmistakable evidence that deeper yet it is hotter still. Eighteenth-century scientists reasoned that the elevated temperature below ground represents a gradual cooling of the planet since its formation, the heat still flowing upwards. They experimented with various-sized spheres of warmed rock and metal, and noted how long it took these to cool, scaling their results up to derive ages for the planet that were much longer than hitherto suspected.

Actually that basic technique is flawed, because it is tacitly assumed that the Earth has no internal heat generation, the temperature differential representing a fossil remnant from the planet's formation as a molten sphere. In fact energy is liberated deep within our globe through radioactive decay; but stepping back a century or two the phenomenon of radioactivity was yet unsuspected. This relates to the problem of the Sun's energy: it was assumed in that era that the Sun was glowing hot because, as a much larger body, it had cooled less than the Earth from its primordial state. The notion of nuclear reactions powering the Sun was unknown until early in the twentieth century.

The whole question was brought to a head when Darwin and his colleagues, studying geological strata such as limestone, showed that sedimentary rock sequences must be hundreds of millions of years old if laid down at a similar rate to those in production today. Up to that point the physicists, on the one side, who were measuring cooling rates and so on, had been able to reconcile their values with the age of the Earth according to geologists, biologists, and the like. However, such a vast planetary and solar age could not be accommodated by the physical theory of the time.

So physicists looked to other possible energy sources for the Sun. If the Sun were gradually shrinking, energy could be produced and the Sun heated. The process may be thought similar to a tennis ball warming as it is compressed whenever struck by a player. During a championship tennis match the balls heat up and this alters their bounce characteristics; cool ones are retrieved from the refrigerator every so often. The familiar phrase "New balls, please" is uttered by the umpire every seven games at Wimbledon. In the case of the Sun or some similar large object, as it contracts there is a decrease in its gravitational energy because the composite matter is moving closer to the middle, and that energy has to go somewhere. Half of it is converted into heat, which is then lost by radiation.

This shrinkage producing heating and hence radiation is a process that is known to occur in the Solar System. Although such a source is insufficient to explain the observed solar power output, we recognize that Jupiter is still settling after its formation so long ago. In consequence it emits two and a half times more energy than it receives from the Sun. Jupiter is not hot enough to emit visible light (we see it only by reflected sunlight), but it does radiate a huge flux of microwaves, making it quite bright to a radio telescope. Saturn and Neptune do likewise, although to lesser extents, whereas the data with respect to Uranus are ambiguous. For the Sun, there is no ambiguity: no such settling could explain the enormous radiated flux of light.

A suggested alternative solar energy source was that meteoroids and other debris continually cascade down upon the Sun; although the individual particles could not be seen burning up, their combined contributions might power the solar furnace. Again, however, the sums would not add up, and the feasible age for the Sun calculated that way was much less than the geologists insisted upon.

A major confrontation over this matter therefore ensued late in the nineteenth century, the physicists seeing a relatively youthful Sun and Earth, the geologists requiring hundreds of millions of years of elapsed time to explain their data. In this argument some physicists acted rather arrogantly, with disregard for what they saw as "softer" scientific disciplines, and yet it was physics itself that threw up the solution and proved these earlier physicists wrong.

MASS–ENERGY EQUIVALENCE

All readers will have heard of Albert Einstein and his Theory of Relativity, but few recognize that there are two rather distinct divisions to it. The so-called "Special" Theory of Relativity is special in that it is limited in scope, whereas the "General" Theory of Relativity is much wider ranging. The latter is often referred to as "GTR" for short, and in essence it may be thought of as being a more sophisticated gravitational theory than that of Newton.

But we must begin with the Special Theory. In 1905 Einstein published four papers on different topics, one of which presented the famous equation showing the equivalence of mass and energy ($E=mc^2$). Here E represents the energy (in Joules), m the mass (in kilograms), and c the speed of light (300 million meters per second). (Einstein actually got his Nobel Prize for one of the other papers, which explained the "photoelectric effect"; his analysis showed that light is split into discrete packets, or *photons*.) Using that equation, and knowing the flux of solar energy at the Earth and our distance from the Sun, it is trivial to show that our local star is losing mass by conversion to energy at an astounding rate, about four million tons per second. Over millions and billions of years it is obvious that the total mass lost must have been enormous, but in terms of the entire bulk of the Sun it is a minor fraction.

The problem of the solar power source was solved, and astronomers at last knew how the Sun and stars shine. From various lines of investigation, especially radioactive dating of terrestrial rocks and meteorites, we now have good reasons to believe that the whole Solar System formed together about 4.5 billion years ago.

The above account glossed over the fact that merely knowing about mass–energy equivalence does not provide an understanding of the complexities of nuclear reactions. Developing such an understanding was the work of many scientists over the subsequent decades. One man in particular, British astrophysicist Arthur Stanley Eddington, was largely responsible for elucidating the physical behavior of stellar interiors in the 1920s.

THE GENERAL THEORY OF RELATIVITY

Eddington had started his astronomical research some years before, in the climate of excitement surrounding Einstein's GTR, which was issued in dribs and drabs before being finalized in 1916. One story often retold is that at a scientific meeting someone mentioned to him that he must be one of only three people who understood relativity, this resulting in Eddington looking puzzled. When chided not to be so modest, his reply was "On the contrary, I am trying to think who the third person might be."

The GTR was viewed as being hugely complicated and disbelieved by many. It presented an entirely new concept of the universe, in which space–time is warped by the presence of matter. This notion always gives trouble to people because they think that their everyday experiences of the physical world can be translated into a comprehension of how the whole universe behaves. This is simply wrong. Einstein's theory was revolutionary in that it said that the shape of space itself is changed by the distribution of matter. This has various concomitant effects, such as clocks going slower (time itself being slowed down) if they are in the proximity of a large mass, or if they are moving through space at a high speed.

If Einstein's theory was to be accepted, it had to demonstrate that it could predict or explain some observed phenomenon when the Newtonian theory could not. It was quickly realized that a previously known anomaly in the orbital motion of Mercury was explicable with the relativistic theory. (This had been a long-standing puzzle, as we will see in Chapter 13.) Einstein's opponents argued that this was a convoluted matter that might be resolved in some other way without recourse to relativity theory. A simpler demonstration of the truth of relativity was required, and Eddington recognized that a total solar eclipse provided a possibility.

THE GREAT ECLIPSE OF 1919

Eddington knew a few things about eclipses (he had gone eclipse chasing to Brazil in 1912 as a member of a large British party which had been clouded out), and he saw how a total solar eclipse could provide a unique opportunity to provide verification for Einstein's theory. The reason for this is illustrated in Figure 4-1.

Consider the light from some distant star passing by the Sun. The path of the light is bent by the Sun's gravity (the rule you may have been taught at school that light travels in straight lines is only a first-order approximation). According to Einstein's theory the bending of the path of the light beam is twice that which Newton's theory of gravity would suggest.

In principle this provides a test, but when one does the sums it turns out that the angles are extremely small. Even for light passing just above the Sun's surface, for which the bending is greatest, the direction change is less than two seconds of arc. How much is that? A degree may be split into 60 minutes of arc, each of

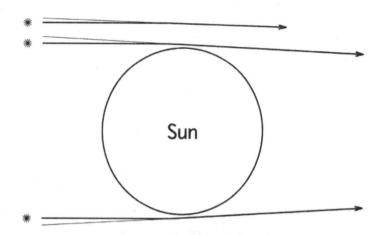

FIGURE 4-1. The deviation of starlight produced by the mass of the Sun, detectable during a total solar eclipse. The paths the light takes from the distant stars at left follow the heavy lines, but from Earth the arrival directions extrapolated backwards appear further from the Sun, as shown by the faint lines. The deflection angles are shown greatly exaggerated. Einstein's relativity theory said that the deflection would be twice that based on Newtonian gravitational theory, and this was verified using the great eclipse of 1919.

which comprises 60 seconds of arc (using the addendum "of arc" to show that we are referring to angles here, not units of time). To put that into some context, two seconds of arc is the apparent width of a matchstick viewed from 220 yards, almost twice the length of a football field. The test would involve being able to differentiate between a single matchstick width, and merely half that as the Newtonian theory would have it.

The problem is that starlight passing so close by the Sun is drowned in the solar glare at all times except during a total eclipse, and so Eddington proposed making observations during such an event. Just any eclipse would not do though. Not only did Eddington need totality, he also needed stars, because the project would not work unless there were several bright stars close to the limb of the Sun during the eclipse. Looking up the eclipse predictions, Eddington saw that one of those represented in Figure 2-2, occurring on May 29, 1919, allowed a unique opportunity. Not only was the totality long, at 6 minutes and 51 seconds, but it was also in late May when the Sun is passing through the constellation Taurus, and crossing a rich cluster of bright stars known as the Hyades.

His mentor, the Astronomer Royal Sir Frank Dyson, was so enthused about the concept that he lobbied the government to avoid having the youthful Eddington drafted to fight in the First World War. Instead Eddington was allowed to prepare for the great eclipse expedition of 1919. The British foray was in several parts, with Eddington leading one group to Principe (a tiny island owned by Portugal, just north of the equator and 150 miles from the African coast), while another headed for the opposite side of the Atlantic, setting up their equipment at Sobral in northeastern Brazil. A contemporary map of the eclipse track, indicating when the footprint reached different locations, is shown in Figure 4-2.

EQUIPMENT CONSIDERATIONS AND PREPARATIONS

If tracks of totality were so considerate as to pass across well-established observatories, then astronomers' lives would be simpler. However, the tracks dictate where one must go to obtain the desired data, which means setting up one-off observatories. The

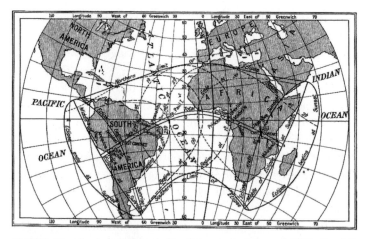

FIGURE 4-2. The track followed by the great eclipse of 1919. The British expeditions were sent to the northeast of Brazil and the island of Principe (labeled here as Princes I).

1919 and other eclipses had to be observed using equipment that perforce was portable, sturdy enough to resist transport to distant spots of high humidity and temperature, and yet easy enough to erect on temporary mounts and then dismantle after use.

In a permanent observatory it is essential that vibrations of the telescope be limited, so that long exposures on faint celestial objects are possible. A telescope is normally bolted to a vast concrete plinth around which the observatory dome can rotate without touching it, and the instrument isolated from tremors. Solar eclipses, though, are relatively brief affairs and so such stability is not as great a problem, making wooden frames like that shown in Figure 4-3 a desirable and practical solution.

A simple mirror to track the Sun, coupled with a long focal

FIGURE 4-3. The wooden mount for a large camera used in photographing the 1922 eclipse from Australia.

length to produce an image as in Figure 4-4, provides an excellent means to photograph an eclipse. Another point to note is that the frame is an open lattice, with only the box around the focus (the actual camera) baffled; if the whole length was enclosed then the Sun's rays would heat it, causing turbulence of the air within and distorting the image.

Although a horizontal arrangement has many advantages, there is a problem with stray light entering the camera. If an eclipse is due on your own doorstep, as was the case in 1918 when

FIGURE 4-4. A heliostat (a rotating mirror used to track the Sun) may be used to reflect an image horizontally into a camera. This is an easier solution than having a long camera tube directed skywards (see Figure 4-5).

a track of totality crossed the United States, one can be a little more extravagant with the preparations. Figure 4-5 shows the scene near the town of Baker, in eastern Oregon, chosen as the best location for observations. The 40-foot-long camera tube was directed towards the precalculated position of the Sun during the eclipse. No matter where the eclipse, clearly setting up the necessary equipment would have posed a major task. Figure 4-6 is a

FIGURE 4-5. The 40-foot camera used to photograph the American eclipse of June 8, 1918.

photograph of a proud array of sailors from the U.S. Navy, plus other helpers, and of course the astronomers, after they had set up the cameras to photograph an eclipse in Spain in the early 1900s.

THE 1919 ECLIPSE RESULTS

In 1919 the British observations did not go smoothly either in Brazil or on Principe, but the altered positions of the target stars were still measurable on the photographic plates exposed.

The astronomers did not immediately break camp and head back to England to announce their results. First they had to wait

FIGURE 4-6. Sailors of the U.S. Navy, having labored to erect the instruments to view a total solar eclipse, pose with astronomers in Spain early in the 1900s.

some months before again photographing the star fields at night, when the Sun was far away, so that the space through which the starlight traveled was not warped by the solar gravity. It was only by directly comparing the two sets of plates that the subtle shifts in the stellar positions would be discernable. They were looking for a differential shift of less than one second of arc; even on a perfectly still night, the amount of scintillation or blurring shown by stars due to atmospheric turbulence is of this order. (Recall the nursery rhyme: "Twinkle, twinkle, little star, how I wonder what you are.")

It was November of 1919 before the outcome of the eclipse analysis was made public, with great fanfare in London. Einstein was right, Dyson and Eddington said, and it was front-page news around the globe.

In subsequent years data collected at other eclipses has clearly confirmed that the deviation of starlight is just as Einstein anticipated. For instance, photographs taken from Mauritania during the great eclipse of 1973 (as plotted in Figure 2-2) again demonstrated that the stellar displacements are larger than Newtonian physics would allow. Measurements using large arrays of radio telescopes have shown that the gravitational deflection of starlight is within one percent of Einstein's value. These and other experiments have shown that Einstein's GTR gives a better representation of the universe than Newton's theory of gravity. Nevertheless, it is possible that there are refinements yet to be discovered.

GRAVITATIONAL LENSES

At the close of the previous chapter I mentioned Charlie Chaplin and his magnum opus *Modern Times*, a movie first shown in 1936. In the same year Albert Einstein published a short note in the journal *Science* concerning how starlight might be focused by gravitational fields. The gist of his paper was as follows.

Consider again Figure 4-1 and imagine the light beams being extended to the right until they meet. Then you could think of the Sun as having acted as a lens: a gravitational lens. Light passing the Sun at top and bottom is brought to a common focus, well off the page compared to the scale of that diagram.

Astronomers like to use big telescopes for two distinct reasons. One is that a larger mirror or lens collects more light, making fainter objects detectable. The other is that better resolution or acuity is, in principle, possible when a large aperture is employed. In reality, however, the turbulence of the terrestrial atmosphere limits the resolution achievable with ground-based telescopes; this movement causes the scintillation (the technical term for twinkling) of stars. This is one of the reasons for putting devices like the Hubble Space Telescope into orbit, above the blurring effect of the atmosphere.

Suppose we positioned a satellite at the focus of the solar gravitational lens, the extended Figure 4-1. With an occulting disk obscuring the Sun, an artificial eclipse would be produced. In a ring around the edge of the disk, the light coming from some hugely distant star or planet would be focused by the solar gravity. The width of the aperture produced by this "solar gravitational lens" would be phenomenal, the Sun being about 865,000 miles in diameter. This would give a resolution—a measure of the smallest detail possible—totally outstripping anything we can achieve either from Earth or using satellites like Hubble.

This solar lens concept all sounds very nice, but is it practical? Actually, when one puts the relevant figures into the equations one calculates a focal length for the solar gravitational lens (the distance to where the lines extrapolated to the right in Figure 4-1 meet each other) of about 500 times the Sun-Earth distance (the *astronomical unit* or *AU*). This would mean that your imaginary satellite would need to be located out beyond all the planets, a dozen times as far away as Pluto. So it doesn't appear to be a feasible proposition, at least within the next several decades.

Might we, though, see the gravitational lens effect produced by some other star? The nearest stars are about 260,000 AU away (this is equivalent to about 4.2 light-years). Other stars have different masses and sizes from the Sun, and so produce all sorts of focal lengths. It could happen that the Solar System is close to the focus produced by some relatively nearby star (nearby on the cosmic scale that is).

This is what Einstein discussed in his 1936 paper: the possibility of other stars producing gravitational lenses. It is a nice idea, but for us to see anything in this way some object of interest must lie very close to the extrapolated line from the Earth to the star acting as a lens, and then beyond, and the probability of such a coincidence occurring is miniscule. For that reason Einstein considered his note only of theoretical interest; "Of course, there is no hope of observing this phenomenon directly," he wrote.

Here, though, the great man's imagination had failed him. He was thinking only of the chance of *individual* stars within our own galaxy, the Milky Way, acting this way. Single stars are of comparatively small mass, cosmically speaking, and so produce little deflection of light beams. Whole galaxies, made up of hundreds of billions of stars, can produce greater effects though. In the 1930s the cosmic distance scale and the characteristics of galaxies were only just beginning to be comprehended, so Einstein can hardly be blamed for his comment. But it was wrong.

A year later another astronomer suggested that galaxies might produce such a lensing effect, but it was four more decades before the first example was uncovered. Several more examples followed, and in the 1990s the search for gravitational lenses became a major pursuit of astronomers, with detection becoming commonplace. The basic idea is shown in Figure 4-7, with interstitial masses such as the spiral galaxy sketched there producing distorted images of

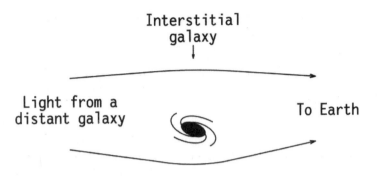

FIGURE 4-7. How an intervening galaxy may cause gravitational lensing of some distant object, producing multiple images that may be amplified in brightness. The deflection angles shown here are greatly exaggerated.

FIGURE 4-8. This image shows the effect of a gravitational lens. The bright smudge in the center is a massive galaxy, but arrayed about it are four other spots, those at the top and bottom being especially bright. These are separate images of some more distant quasar, focused by the lens action of the intervening object. In essence the galaxy is eclipsing the quasar, but paradoxically its gravitational lens effect brightens the light received from the latter. These quadruple images form what is known as an Einstein Cross. It is also possible for other slight misalignments to produce bright images that are double, triple, arcuate, or other distorted forms. In the case of a precise alignment, a circular image, called an Einstein Ring, is formed.

FIGURE 4-9. This Hubble Space Telescope image shows the gravitational focusing effect of a huge cluster of galaxies known as Abell 2218. The many arcs spread across the photograph are distorted images of other galaxies and quasars five to ten times as far away from us as the cluster causing the lensing.

more-distant light sources. An example is shown in Figure 4-8, a focusing galaxy producing four images of a distant *quasar* (that is, a quasi-stellar object; the true nature of such sources is still unknown, but they seem to be very distant but extremely luminous objects). The effect is similar to that obtained by looking through the bottom of a wineglass, where a variety of distorted images form as you move your eye around. This is more obvious in Figure 4-9, a photograph of a cluster of galaxies whose combined gravity leads to many arcuate images of more distant galaxies and quasars that cannot otherwise be seen.

Is the phenomenon seen in Figure 4-8 an eclipse? Yes, because the focussing galaxy is blocking our direct view of what is behind it. Paradoxically the eclipse is amplifying the brightness of the quasar, in the same way a magnifying glass enhances the intensity of sunlight such that a piece of paper may be ignited. Without that amplification the quasar might well have been too faint to detect, so that, rather than simply hiding it, the galactic eclipse has made possible the detection of this light source at the periphery of the universe.

It happens that Einstein was also wrong in the context of observing gravitational lenses within the Milky Way. The gravitational lens formed by a single star is extremely narrow, and so even with about 400 billion stars in our galaxy he reasoned that the chance of getting two stars aligned with the Earth at the focal position must be exceedingly small. That is based on the assumption of a static situation though. In fact, all the stars are moving, in orbit around the galactic center and also shifting relative to each other with their own peculiar velocity components. Every so often two *will* align with the Earth, and major astronomical research projects now automatically monitor thousands of stars each night. As an alignment occurs, the focussing through the transient "lens" produces an increase in intensity (like the magnifying glass again), and this brightening may persist for days or months. The computers scanning the images are programmed to draw attention to such intensity enhancements. Using the results, astronomers are also searching for the so-called "missing mass" that seems to hold our universe together.

5

The Turbulent Sun

A few seconds before the Sun was all hid, there discovered itself round the Moon a luminous ring about a digit, or perhaps a tenth part of the Moon's diameter, in breadth. It was of a pale whiteness, or rather pearl-colour, seeming to me a little tinged with the colors of the iris, and to be concentric with the Moon.

Edmond Halley (1715), describing his observation of the corona, which he took to be of lunar rather than solar origin

The Sun is of huge importance to life on Earth, making it very special. Nevertheless, leaving aside our natural bias one has to say that it is not special at all when compared with other stars. There are reckoned to be about 400 billion stars in our galaxy, the Milky Way. There are blue-white supergiants, brown dwarfs, pulsars or neutron stars, white dwarfs, red giants, black holes, binary stars we know to be double only from their spectra, X-ray emitting binaries, and too many other distinct categories of stellar creature to mention, let alone describe their properties.

Most stars are rather nondescript, spending most of their lives on what is termed the *main sequence*, an evolutionary track along which stars with different masses, ages, and chemical compositions are burning hydrogen within their cores. ("Burning" here does not mean simple combustion, which is a chemical reaction with oxygen, but rather *nuclear* burning, in that hydrogen nuclei join together to produce helium.) As they do so, they generate far more energy than any trivial chemical reaction, just as nuclear bombs liberate rather more energy than dynamite.

Thankfully our Sun is one of these nondescript stars. Our neighborhood nuclear generator behaves in a regular way, not burping out vast quantities of its star stuff and incinerating any nearby planets, nor shrinking to leave its rocky companions to a frigid existence. At least, the Sun will not do so yet. It has been merrily emitting energy generated by those nuclear reactions in its core for about 4.5 billion years. It is expected to do the same for another 5 to 10 billion before swelling up into a red giant, enveloping the planets and asteroids out as far as Jupiter, and then collapsing into a white dwarf, having exhausted its nuclear fuel. As it shrinks it may cast off a nebula of gas and dust, which would eventually be recycled to help produce yet more stars and planets.

Some other stars are massive enough such that their cores attain pressures and temperatures sufficient to burn heavier elements, like carbon and nitrogen, producing elements with ever more particles in their nuclei, and so extending the stellar lifetimes. But our Sun cannot do so. Its lifetime is limited. Let us not weep, though: if the Sun were not *just* as it is, we would not be here to appreciate it and grieve for its eventual expiration.

INSIDE THE SUN

The Sun contains 99.8 percent of the Solar System's mass (most of the rest of it is in Jupiter), about 330,000 times the bulk of the Earth. Around 73 percent of the Sun is hydrogen, 25 percent is helium, and all the other elements added together comprise less than 2 percent of the solar mass.

The Sun agglomerated from a huge cloud of gas and dust, which was largely the debris left from previous expired stars and supernova explosions. In its core the temperature is over 10 million degrees Celsius (20 million degrees Fahrenheit), and the pressure is in excess of 200 billion times our atmospheric pressure. We say that the material within the Sun is a gas, and yet its density is 150 times that of water, 20-fold that of iron.

Under such conditions the repulsive forces between hydrogen nuclei may be overcome. (Hydrogen nuclei are simply bare protons—positively charged subatomic particles, the number of which within any nucleus controls the type of element it is.) Helium is produced as they coalesce. That is, the Sun is a natural fusion reactor. If we could do the same thing on Earth we would have a practically unlimited supply of energy, although one could not say that it would be *free* because many, many billions of dollars have already been spent in the as yet unsuccessful quest to produce controllable fusion. (Uncontrolled fusion is easy: it's called a hydrogen bomb.)

Since the time this fusion process began in the center of the Sun just over four and a half billion years ago, about half of the usable hydrogen fuel has been transmuted into helium. The word "usable" is significant here because, as the hydrogen at the middle is consumed, the shell where fusion is occurring moves outwards. But away from the center the temperatures and pressures eventually become too low to support hydrogen burning, and so fusion halts. This means that much of the hydrogen in the Sun will never be burnt. If the interior of the Sun were better mixed then it might have a longer lifetime, but things are as they are, and stellar interiors are heavily stratified.

Figure 5-1 shows a schematic cross section of the Sun. Energy generation through fusion occurs only in the core, which occupies about 25 percent of the overall radius. That energy is transported outwards through the *radiative zone*, the next 50 percent or more of the radius. The energy is carried through that zone by photons of light, rather than by conduction or convection. (Conduction is the process of hot atoms colliding with cooler ones and transporting heat away, in the same way as the handle of a long iron rod gets warm if the other end is left in a fire. Convection is

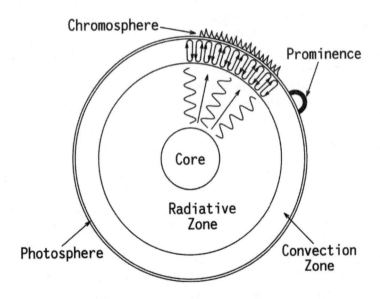

FIGURE 5-1. A cross section through the Sun showing its basic features at different levels.

the wholesale upward movement of hot atoms, like heated air rising above a stove.) Those photons deep within the Sun are not at the wavelengths of visible light; the temperatures there are so high that the photons are mainly in the gamma- and X-ray region of the spectrum.

The final 20 percent of the solar radius is known as the *convection zone*. In this layer the temperature gradient is sufficient for bubbles of hot sun-stuff to rise until close to the surface, giving the Sun its characteristic mottled appearance. (The effect is similar to making gravy or jam: the heating at the base of the pan makes the liquid want to rise, but not all of the liquid can rise at once, so it churns over in globules moving together.) After cooling, that material sinks again to the base of the convection zone, where it is heated again before beginning another round trip to the surface, as part of another cell. This convection results in the Sun's surface not being smooth, but covered with thousands of these globules, which are called *granules*. They are each the size of a continent, but short-lived, persisting for but a few minutes before dissipating and being replaced by some new rising globule.

What is usually referred to as the "surface" of the Sun is correctly termed the *photosphere* (that is, the region from which our eyes detect photons). This is not a solid surface, but a layer of ionized gas at a temperature of about 5,700 degrees Celsius (10,300 degrees Fahrenheit). The temperature of the photosphere determines the color we perceive: that's why the Sun appears yellow to us, whereas hotter stars appear blue or white, and cooler ones orange or red.

In Figure 1-1 we saw photographs of the Sun's surface, including some sunspots. These are cooler regions of the photosphere, typically at around 4,000 degrees Celsius (7,200 degrees

Fahrenheit). Their origin is not yet completely understood, although they are certainly related to convolutions of the intense solar magnetic field. In a sunspot the magnetic field is several thousand times as intense as elsewhere on the solar surface. One should not underestimate their size: many are 25,000 miles across, several times the diameter of the Earth.

OUTSIDE THE SUN

Above the photosphere is an almost translucent region known as the *chromosphere* due to its scarlet coloration. This color results from its hydrogen content, which emits visible radiation largely at a specific red wavelength. The chromosphere is quite thin: a few thousand miles wide, which is large on the scale of a planet, but less than 1 percent of the solar diameter.

Penetrating the chromosphere are spikes of gas that rapidly jet upwards and then fall back again; these are termed *spicules*. Larger ejections of mass are called *prominences*, as seen in Figures 1-3 to 1-5. Such prominences provide one of the highlights of a total eclipse.

Another vivid feature seen in an eclipse is the *corona* (or *aureola*). This is a rarified region of extremely hot gas, stretching millions of miles out into space, consisting of ionized atoms speeding away from the Sun. Like the chromosphere, the corona can only be seen by eye during a total eclipse, although there are other technical ways to observe it between times. One of the great puzzles of solar physics is how the corona is heated to such a high temperature—over a million degrees Celsius (2 million degrees Fahrenheit)—given that the underlying regions are much cooler.

Flowing outwards from the Sun is a continuous stream of

particles known as the *solar wind*. These particles zip by the Earth at a speed of about 300 miles per second. The Sun has an intense but dynamically changing magnetic field that is carried outwards by the solar wind. By dint of their own magnetic fields, the planets interact with this solar wind, producing effects both beautiful, like the auroras, and disruptive, such as interference with radio communications, navigation systems, TV and cell phone services, and manned space walks. The density and other characteristics of the solar wind are quite variable. For example, there are gradual ebbs and flows with the 11-year solar cycle, but also spasmodic *solar flares* may be seen, associated with ejections of large amounts of matter into the solar wind, intersecting the Earth a day or two later.

The photographs in Figure 5-2, which were obtained using an instrument known as a coronagraph (see below), show the corona and solar wind, both of which are highly uneven. Coronal streamers are the most obvious features in those images, their shapes varying over the several hours of the data collection.

ARTIFICIAL ECLIPSES

There are many aspects of the Sun and interplanetary space that may only be studied during a solar eclipse, and yet such a natural event occurs only every year and a half or so, and often then in inhospitable places from the perspective of astronomical observations. It is therefore natural to wonder whether it is feasible to create an artificial eclipse, by using a circular baffle to imitate the action of the Moon when it passes in front of our nearby star.

A telescope designed to do this is called a *coronagraph*; that is, it is used to study the corona. It is equipped with an obscuration to

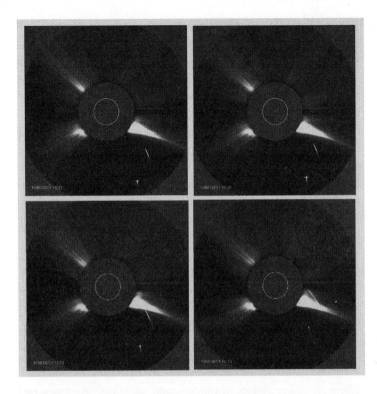

FIGURE 5-2. This four-hour sequence of images obtained with the coronagraph on the SOHO satellite (see page 133) shows two comets falling into the Sun in June 1998. Several distinct coronal streamers are obvious in these images, their forms altering over these several hours. A coronagraph produces an artificial eclipse, allowing the solar atmosphere to be studied continually, rather than only during the few minutes of a natural eclipse. (There are many other advantages of data collection from space, such as access to wavelengths that are absorbed by the terrestrial atmosphere. An example is the ultraviolet image shown in Figure 1-5.)

fit over the image of the luminous solar disk. Astronomers have been using coronagraphs for some time, but they are of limited utility down here on Earth because the atmosphere scatters so much sunlight. The innermost corona is about one-millionth the brightness of the photosphere. Stepping out by a distance equivalent to the solar radius, the coronal brightness drops by about a factor of a thousand, to be about a billionth that of the solar disk. The light of the sky is greater than this value, meaning that the studies that may be tackled using ground-based coronagraphs are limited. By launching a coronagraph on a satellite, however, astronomers can get their instrument above the atmosphere and avoid such drawbacks. It just costs a lot more. Far above the Earth, a coronagraph can be guided so as to keep its baffle over the Sun, allowing perpetual monitoring of the corona, prominences, and other solar phenomena. Over the past couple of decades several such telescopes have been launched, returning invaluable data.

The most advanced satellite of the type is the Solar and Heliospheric Observatory (or SOHO), a joint project of NASA and the European Space Agency. On board are 11 separate instruments, one of which is called LASCO (Large Angle Spectroscopic Coronagraph). This has allowed the experimenters to study the corona out to 30 times the radius of the Sun, watching how vast bodies of hot plasma (bubbles of highly ionized, or charged, gases) are thrown out into space. Such spasmodic events, termed *coronal mass ejections*, sometimes result in a hundred billion tons of sunstuff being launched outwards. These disperse somewhat as they move away from the Sun, but they can affect us on the Earth in various ways, because our ionosphere (the layer of the upper atmosphere consisting of gases ionized largely by the solar ultraviolet radiation) is disturbed as we run through streams of this plasma.

FIGURE 5-3. This image obtained by the *Clementine* satellite in 1994 shows the solar corona shining above the limb of the Moon. The bright disk is Venus. The part of the Moon towards the spacecraft obviously cannot be illuminated directly by the Sun, as that is on the far side of the Moon. In fact the lunar surface can be seen clearly here owing to Earthshine: sunlight that has been reflected by the Earth, which is off the field of this image to the right.

These types of solar gas ejection can be seen in Figure 5-2. The frames shown there also display a quite distinct phenomenon: comets falling into the Sun.

Another way to achieve a type of artificial eclipse is to use the Moon as the baffle and fly a satellite into its shadow. An example of this is shown in Figure 5-3, which is an image obtained by the *Clementine* spacecraft in 1994 while it was in orbit around the Moon.

THE HEYDAY OF SOLAR ECLIPSE CHASING

It is easy to give a verbal sketch of the basic features of the Sun, as above, and forget that our understanding has been gradually developed over many years. That development has certainly accelerated during the Space Age by a huge factor, but we should not decry the forward steps taken in earlier times. Indeed the lines of thinking in different epochs provide instructive examples of the science of the day. Let us begin by noting that the effort and expenditure that go into solar eclipse watching has diminished over the last several decades and examine why that is so.

Although many enthusiasts venture around the globe for each total eclipse, and still there is much useful professional research that can be done, the heyday of eclipse watching was between the 1840s and the 1930s. An eclipse sweeping over southern Europe in 1842 was witnessed by many. Francis Baily observed it from Italy, confirming his impression of the eponymous "bright beads" that he sketched in 1836 (refer back to Figure 1-7). Those transient baubles of brightness are produced by sunlight propagating to one's eye betwixt and between the craggy mountains and crater rims at the edge of the Moon.

Many important discoveries concerning the Sun's properties were made in those days, and we will meet them below, but for the present we should just recognize the extraordinary fervor with which eclipses were chased through that era. In 1870, for example, French astronomer Jules Janssen was so desperate to get to Algeria to observe an eclipse that he escaped from Paris in a balloon, drifting over the heads of the Prussian troops who had the city under siege.

Moving forward a few decades, governments were prepared to provide expeditions to eclipses with financial support orders of magnitude higher than would be conceivable nowadays (see Figures 5-4 and 5-5). Today astronomers might be pushed to raise the necessary capital to cover economy-class airfares for themselves

FIGURE 5-4. The epic days of eclipse expeditions: this is the British team in Imperial India, at Baikal on the southwest coast of the sub-continent, waiting for the eclipse in December 1871.

and a couple of assistants to take their instruments to some eclipse track. Contrast that with the cost to the British government of staging the 1919 eclipse expeditions, large teams spending several months in both Principe and Brazil.

Those days are over. Radio astronomy began as a science after the Second World War, following the first stuttering observations in the 1930s. Similarly the start of the Space Age in the late 1950s, with its blossoming since, has opened up new areas of research

FIGURE 5-5. An eclipse team at drill before the big event, located near the northern tip of Scandinavia in 1896.

capability, and these have lessened the scientific significance of natural eclipses. With satellite instruments we are now able to produce artificial eclipses at will. One should, however, pay proper regard to the pioneering eclipse observations made over the past centuries, and how they enabled many of the properties of the Sun to be elucidated.

UNDERSTANDING THE CORONA

Nineteenth-century astrophysicists were much confused about the source of the Sun's power. They did not understand this until nuclear reactions were discovered, as discussed in Chapter 4. This had major ramifications for other areas of science because the origin of solar energy affected estimates of the age of the Earth, and hence studies of geological and biological evolution. In this earlier state of ignorance they thought of the Sun as burning in the same way as does wood or coal, combustion through a chemical reaction with oxygen. Thinking in that manner they were bound to interpret certain phenomena in erroneous ways.

Take the prominences, for example. If you look up a nine-teenth-century book describing the Sun, these will often be called simply the "red flames," and flames are what they were commonly thought to be, licking upwards from the solar surface like a huge spherical bonfire. One could then imagine that the inside of the Sun would be cool, because the burning had not yet penetrated there, just as a charred piece of wood is neither hot nor burnt in its middle. Some thought that sunspots were holes through this supposed layer of burning. Sir William Herschel, who discovered the planet Uranus from the city of Bath in England in 1781, opined that there might be alien beings living inside the Sun, down below the burning layer, where the conditions were wrongly imagined to be pleasant.

Similarly the nature of the corona was only gradually understood. Its existence was well known to the ancients. Plutarch wrote, apparently in reference to an eclipse he had witnessed from Greece in A.D. 70, that "Even if the Moon, however, does sometimes cover the Sun entirely . . . a kind of light is visible about the rim which keeps the shadow from being profound and absolute." The first description of the corona in modern astronomy was by Johann Kepler, who described its appearance during an eclipse over Prague in 1605. Giovanni Cassini, of the Paris Observatory, made a more complete identification in 1706.

These early observers were not sure whether the corona was a solar or a lunar phenomenon: was it perhaps a lunar atmosphere which could only be seen when the Sun was behind the Moon, suitably illuminating it? This is what Edmond Halley believed, as exemplified in the quotation that heads this chapter. In 1715 he suggested that the apparent asymmetry of the corona was due to the Sun heating only one face of the Moon at any time, building up a gaseous cloud above that hemisphere that would condense as the Moon turned, like the diurnal cycle of dew.

The fact that the corona is the extended *solar* atmosphere was not settled until 1890. Much later other components of the detected light were identified, such as the *F-corona*: sunlight scattered by a sphere of dust grains stretching out to tens of solar radii. This is similar to a tenuous dust cloud on Earth, visible only from the light it scatters or absorbs.

How can the gaseous corona be investigated? One of the fundamental techniques used in astronomy is spectroscopy: the study of the spectra produced by different sources of light. The first step along this path was taken in 1664 when Isaac Newton used a prism to split sunlight into its constituent colors, the familiar rainbow. In the early nineteenth century the great German physicist Joseph von Fraunhofer, using sophisticated optical devices to disperse the light more widely, showed that sunlight is not an unbroken spectrum: at certain wavelengths there are dark bands. With excellent resolution, thousands of these may be identified. Their origin is as follows.

The photosphere, at a temperature of several thousand degrees, emits a continuous spectrum (that is, all wavelengths), just as an electric light globe does. The tenuous upper layers of the Sun are cooler and tend to absorb light. They do not do so over the whole spectrum, but only at distinct wavelengths, the precise character of which depends on the chemical elements present. That is, iron will absorb at one set of wavelengths, chromium at another, carbon at another, and so on. Therefore each element produces its own characteristic *absorption spectrum*: the continuous spectrum emitted by the hot gas below will be modified such that there are many dark spectral lines crossing it, where cooler gas higher up has absorbed certain wavelengths. This is of great practical importance because by studying these absorption lines astronomers can gauge the quantities, temperatures, and ionization states of not only the constituents of the Sun, but also other stars or the atmospheres of planets. To do this they use a device known as a *spectrometer* or a *spectroscope* (see Figure 5-6). Similar the ozone layer in the terrestrial atmosphere, and other components of it, can be remotely sensed by spectroscopic means.

If the atoms are hot then an element's spectrum will consist of *bright* (or *emission*) *lines*. For example, street lamps emit only particular wavelengths of light: yellow lamps employ sodium, blue-white ones use mercury, and red-strip lamps contain neon.

FIGURE 5-6. A spectroscope of the type used by nineteenth-century astronomers to identify the chemical elements in the Sun, including the discovery of the existence of helium.

The spectrum that is detected, whether emission or absorption, allows the astronomer to ascertain the chemical composition of distant light sources without needing to grab samples and bring them back for laboratory analysis. In the laboratory they can electrically excite, say, calcium atoms in a vacuum tube and measure the wavelengths emitted. Detecting the same wavelength pattern from some astronomical object, they will know that the source also contains calcium. This means that we can identify elements previously known on Earth in the composition of distant stars. But what if the astronomer detects spectral lines that are unknown to science?

THE DISCOVERY OF HELIUM

To the Romans the god of the Sun was *Sol*. To the Greeks, he was *Helios*. The ancient Greeks were more proficient in science and mathematics than the Romans, which is why Greek words are often employed in scientific matters (a term like "heliocentric," for example, or "telescope" from the Greek "tele" meaning "distant").

The element helium gets its name in the same way. This is the second member of the periodic table of elements (the sequence of naturally occurring atoms). It is always found as a monatomic gas—that is, a molecule containing just one atom—because it is inert, meaning that it does not undergo any chemical reactions with other atoms. As a consequence, although it occurs on the Earth its existence had escaped the notice of science until being identified as a major constituent of the Sun—using an eclipse, of course.

The story is quite peculiar. We discussed the so-called Fraunhofer spectrum of the Sun earlier. This consists of the continuum from the lower photosphere, superimposed on which are the many dark lines produced by the cooler atoms in its upper-most layer absorbing at their specific wavelengths. This absorption spectrum can be detected at any time; its intensity swamps any other solar light except in a total solar eclipse. During such an eclipse we see the corona and other structures that are normally drowned by the photosphere. (Similarly, as you look out the window of a railway carriage on a clear day you see the countryside whizzing past. When making the same trip at night, though, or when passing through a tunnel, you see mainly your own reflection and the interior of the carriage from the inside of the glass window. That reflection is always there, but it is not easy to see in broad daylight.)

Turning a spectroscope upon the corona, nineteenth-century astronomers found that the spectrum they recorded was quite un-like the familiar Fraunhofer spectrum. There was no sign of the dark lines, but the exact opposite: all they detected was a series of bright lines. An example of this emission spectrum is shown in Figure 5-7. This is a spectrum of the chromosphere and corona, displaying a series of bright lines produced by the hot gases just above the photosphere. Many of these lines could be identified with known elements, and in particular the red coloration of the chromosphere was recognized to be due to the strong "Hα" line (seen at far right in Figure 5-7). This allowed hydrogen to be identified as the major constituent of the solar atmosphere. (H is the chemical symbol for hydrogen; the Greek letters applied here are simply used by convention to label the specific spectral lines.) But two other distinct lines in Figure 5-7 had an unknown origin.

In 1868 eclipse observations allowed the wavelength of the line immediately to the left of the Hα line (labeled "5875") to be

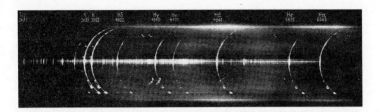

FIGURE 5-7. A spectrum of the chromosphere photographed by nineteenth-century astronomers, showing some of the features that perplexed them. The spectrum was obtained by dispersing the light received during a total solar eclipse using a large prism. The shape of the chromosphere, around the limb of the Sun, is obvious, as are several prominences. At the top various spectral lines are labeled. Five hydrogen lines are labeled (Hα, β, γ, δ, plus another denoted simply by an H at wavelength 3968 angstroms), two of helium (He), one of potassium (K), and one of titanium (Ti). Many other fainter lines are left unlabeled, but are identifiable.

measured accurately. It was realized that this could not possibly be due to sodium (the yellow of sodium street lamps is due to two very close lines at a shade longer wavelength than this). Two astronomers realized concurrently that this was evidence of a previously unknown element: the aforementioned Jules Janssen and British scientist Sir Norman Lockyer. It was Lockyer, who later made a name for himself with his astronomical theories about megalithic monuments and the Egyptian pyramids, who suggested the name helium for the new element, and hence the symbol "He" used in Figure 5-7. Because it is chemically inert, helium was not identified on Earth until some time later, in 1895.

CORONIUM AND NEBULIUM

This is not the end of the story of the coronal spectrum. As seen in Figure 5-7, there are myriad spectral lines, each of which may have its wavelength measured, and its elemental parentage perhaps allotted. Inspired by the discovery of Janssen and Lockyer in the previous year, astronomers flocked to observe the 1869 total eclipse in North America using their spectrometers, and the new methods of photography to record the spectra for later analysis.

When the dust settled and all the "easy" spectral lines had been accounted for, still there were many that could not be ascribed to any known element. A novel species was invented to explain these, and it was called "coronium" because it was found only in the solar corona. Astronomers also turned their telescopes towards the distant nebulas of the cosmos, and found evidence, they thought, for yet another element. This was christened "nebulium."

Coronium and nebulium were both, in the event, figments of the astronomers' imaginations. The lines they detected were real, but their interpretation was wrong. It is possible to get known elements to produce those spectral lines if their atoms are subjected to extreme physical conditions, such as the huge temperatures of the solar corona. Physicists could not produce a temperature of a million degrees in their laboratories, and so these lines had not been seen previously.

When one supplies an atom with some energy, by heating it or by illuminating it with light of a wavelength below some threshold, an electron (a negatively charged particle) can be ejected, leaving an *ion*: a positively charged atom. It is possible to strip off another electron, making the ion doubly charged, and maybe another, but it gets progressively more difficult to remove extra electrons. That limits what can be done on Earth (at least in a controlled way: a nuclear explosion is another matter).

In the solar corona, however, the phenomenal temperatures mean that the ions are multiply charged. As each successive electron is removed, the resulting ion produces a new, distinctive set of spectral lines. For example, greatly ionized iron atoms may have lost ten electrons and emit a series of wavelengths that one could not hope to duplicate in a laboratory. No wonder the astronomers were confused.

How then can we identify the atom responsible? The answer comes from theoretical calculations, although again there is a twist to the tale. There are simple selection rules that usually work in spectroscopy, corresponding to the known properties of atoms. According to these rules, many of the lines detected appeared to correspond to "prohibited" transitions. Such "forbidden lines" were not fully understood until after the developments in quantum theory that took place in the 1930s. From the correct identification of the "coronium lines" it was eventually inferred that the corona is exceedingly hot: over a million degrees, as we saw earlier.

Before leaving coronal spectroscopy, consider an interesting coincidence. In Figure 5-7 the numbers give the wavelengths of various lines in angstroms (one angstrom, which is given the symbol Å, equals one ten-billionth, or 10^{-10}, of a meter). The spectrum we see with our eyes extends from about 4,000 Å (the violet/blue end) through to 7,000 Å (the red end). (Some physicists like to use angstroms for wavelengths, while others use the strict metric system, so you will also find wavelengths given in nanometers. One nanometer [1 nm] is a billionth, or 10^{-9}, of a meter, and so equals ten angstroms.) The angstrom unit gets its name from Anders Ångström (pronounced *ong-struh-m*), a Swedish astronomer who lived from 1814 to 1874. The coincidence is that it was he who first identified the hydrogen lines in the solar spectrum, showing us why the chromosphere is red.

OBSERVING THE CORONA WITHOUT AN ECLIPSE

Once astronomers had understood the basics of the solar spectrum, through the forward leaps in knowledge in the 1860s, it became possible to observe the corona at times other than during a total solar eclipse, by using a suitable filter.

We have already seen how the photosphere produces a continuous spectrum, which is modified by absorption from cooler overlying gases. Consider the red Hα line at 6,563 Å. At that specific wavelength the cool hydrogen at the top of the photosphere absorbs much of the light flux. But the hotter hydrogen in the chromosphere and corona above it is madly emitting at the same wavelength. If one observed the Sun using a filter that lets through only light within a narrow band about 6,563 Å, then much of the photospheric spectrum would be cut out; what comes through would be the emission from these higher reaches of the solar atmosphere. The solar disk is therefore being blocked not by the opaque Moon, as in an eclipse, but by the clever use of a spectral filter. (It is called an "Hα" filter because it blocks all light *except* the wavelength corresponding to the Hα hydrogen spectral line.)

Astronomers soon seized upon this, and from the 1870s onwards it has been a fundamental technique allowing the changing form of the chromosphere to be followed. If you are ever in the presence of a group of astronomers, among the jargon bandied about the term "aitch-alpha filter" will often be heard.

6

Ancient Eclipses and the
Length of the Day

. . . time that takes survey of all the world . . .
William Shakespeare, *Henry IV Part I*

I t may seem surprising, in the Space Age with cosmic phenomena being studied using hugely sophisticated instrumentation,
that relatively crude ancient eclipse records are invaluable, even
irreplaceable, to modern science.

Let me give an example of the value of such eclipse records.
Imagine you suspect the day is getting longer, because the rate of
spin of the Earth is very gradually slowing. You can measure that
spin rate directly in the short term using a host of high technology
equipment: vast arrays of radio telescopes following the motion of
extra-galactic objects across the sky, laser beams reflected from
orbiting satellites, phenomenally precise clocks employing beams
of cesium atoms or hydrogen masers. Data collected using such
techniques indicate that the duration of a day in 2001 was about
0.17 milliseconds longer than it was back in 1991. It's a small
change, but a decade is only a short interval, historically speaking.

Alternatively, one can investigate how the day length has
changed not just over the past decade, but also over 200 or 300

decades. How is this possible? In the first millennium B.C. the Egyptians, Babylonians, and Chinese did not have atomic clocks. In fact they had no artificial clocks at all, apart from simple devices measuring water flow, which were hardly very precise. But they did have natural clocks provided by the Sun and the Moon in the sky.

Suppose that a total solar eclipse was observed and recorded from Athens in 500 B.C., on the local calendar scheme in use in that era. Such eclipses are so infrequent that we can identify the event using our knowledge of the apparent orbits of Sun and Moon about the Earth. The bare observation of the total eclipse tells you that on that date the Sun, Moon, and Athens were aligned (to within a tolerance equal to the width of the eclipse track, which is equivalent to a few minutes of time). This then tells you the local solar time for Athens in that era: that is, when the Sun rose, when it crossed the meridian, when it set.

Since 500 B.C. the day has continually been getting longer. Over a single century the day increases by about 1.7 milliseconds (although there are reasons to believe that this deceleration is variable). This tiny amount summed over 2,500 years gives a total shift amounting to about four hours, equivalent to one-sixth of a rotation of the planet. So, if the day length had stayed the same, the eclipse track would have been out over the Atlantic and would have escaped detection by the Greeks.

The mere recording of an eclipse from Athens so long ago would provide rather accurate information about how our rotation rate has slowed, without the Greeks having made any sophisticated scientific measurements. Actually, no such eclipse occurred at that place and time (I made it up as a thought experiment). However, there *are* records of a similar nature written down by disparate civilizations over the last three millennia. Despite the fact that the ancients had no lasers, artificial satellites, radio telescopes, or cesium clocks, their accounts of eclipses *have* made it possible to build up a consistent picture of how the length of the day has changed.

Not only is the day getting longer (making necessary the insertion of leap seconds), but so, too, is the month, because the Moon is slowly receding from the Earth. In this chapter we consider the implications of these trends.

THE INCREASING DISTANCE TO THE MOON

When, back in Chapter 2, we looked at the fundamental processes by which eclipses eventuate, we noted that the angular diameter of the Moon as seen from the Earth is almost precisely the same as that of the Sun. This is a quite remarkable coincidence. There are small cyclic variations in those apparent sizes because the Earth–Moon distance changes as the latter moves between perigee and apogee, and the Earth–Sun separation alters as the former moves between perihelion and aphelion. Nevertheless it seems a staggering coincidence that the angular diameters of the Sun and the Moon are so similar.

If the dimensions of either Moon or Sun were a little bit different, then the stringent eclipse conditions would collapse. If the Moon were slightly further away then no total solar eclipse could ever occur. Conversely, if it were slightly closer then eclipses would occur more frequently, and we would have added opportunities to wonder at them.

In fact the Moon *was* closer to us in the past. And if you happen to read these words precisely one year after I typed them, then, the Moon will have receded from the Earth by about an inch and a half.

The value I have just given for the increasing separation is derived directly from lunar laser ranging experiments. Between 1969 and 1972 the Apollo astronauts left several retro-reflectors on the lunar surface, these acting similarly to the glass "cat's eyes" inserted along the central line of a road, reflecting back the light from an advancing car's headlamps. The retro-reflectors installed on the lunar surface are similar devices, although rather more sophisticated, shaped like the corner of a cubic prism (see Figure 6-1).

FIGURE 6-1. The laser retro-reflector array left on the surface of the Moon in the Apollo 14 mission in 1971. A hundred separate corner-cube prisms are used in this device. Over the past 30 years it has been used to reflect laser pulses back to observatories on Earth, making it possible to monitor the slow drift of the Moon away from our planet.

By firing short laser pulses towards these through a large telescope and measuring the time it takes for a few of the transmitted photons to be reflected back to the Earth, physicists can measure the round-trip time: slightly over two and a half seconds. Knowing the speed of light, the experimenters can determine rather accurately the current distance to the Moon. Over three decades of such trials they have shown that the recessional speed of the Moon is about an inch and a half per year.

THE NEED FOR LEAP SECONDS

The above conclusion was not unexpected. We already knew the Moon to be drifting away from us very slowly. The Moon raises tides in the oceans, and these create a drag force that is incessantly dropping the terrestrial rotation rate. Although the effect is small, it is both calculable and observable, for example through radio astronomical observations of distant quasars (these are so far away that they provide unmoving references against which the terrestrial spin may be gauged). On top of this persistent slowdown trend, the rotation rate of the planet is also found to undergo seasonal variations, as the atmosphere swells under summer heating and then shrinks in the winter.

It is because of this general slowing down of the Earth that leap seconds need to be inserted into some years. In the past, time was defined astronomically, from observations of when the Sun and the stars crossed the noon meridian. However, during the twentieth century methods of time determination that were of ever increasing accuracy were developed, eventually resulting in time according to the heavens being abandoned in favor of time according to atomic clocks. The atomic second is the standard we use now, and that is defined according to the length of the day as it was in 1900. Over the century that has elapsed since then, the days have become about 1.7 milliseconds longer. Such differences accumulate to give a discrepancy of one second over 19 or 20 months, making a leap second necessary to keep the time shown by atomic clocks in accord with the spin of the planet. Leap seconds are inserted on an as-needed basis, by international agreement, at the end of either December 31 or June 30.

As the years pass the day is getting longer and longer, and leap seconds will eventually be required more often. If the present rules are maintained then within a few centuries we may need a leap second at the end of every month. One way to avoid this would be to redefine the atomic second in terms of the day length in A.D. 2000 rather than 1900, and then no leap seconds would be needed for some decades, but there are problems with such a solution. For example the fundamental unit of length used in all science and technology, the meter, is now stipulated in terms of how far light travels in a second, and so amending the second would alter the definition of the meter. Also, radio frequencies are given in units of Hertz, or cycles per second, so that changing the second would affect those too.

WHY THE MOON IS RECEDING

As the Earth's speed of rotation diminishes owing to tidal friction, its angular momentum falls. The angular momentum of a body is a measure of its disinclination to stop rotating (or indeed to speed up), whether that rotation is in the form of spinning on its axis, or revolving around another body. An example of the latter is any planet orbiting the Sun, or the Moon orbiting the Earth. A body's angular momentum depends upon its total mass, the distribution of that mass, and its rotation rate. The total angular momentum of any system is a quantity that is absolutely conserved (remains the same). An example is a pirouetting ice-skater. With arms outstretched her spin rate may be slow, but as she draws her arms down to her sides, the rate of spin increases. Because the mass distribution has been changed, the spin rate alters to compensate and thus keep the angular momentum constant.

In the case of the Earth there is braking due to tidal drag, because the continents prohibit the free movement of the tidal swell right around the globe, and in consequence the planet's spin angular momentum reduces. We have just seen, though, that the angular momentum *must* remain the same. Evidently, something else has to be happening here if the laws of physics are to be obeyed.

How is this achieved? I noted above that it is the total angular momentum of the *system* that is conserved. Here we are considering the Earth–Moon system as a whole. As the spin angular momentum of the former drops, the angular momentum associated with the orbit of the latter must increase. To make this happen, the Moon recedes from our planet, very slowly. And that is why the lunar laser ranging experiments indicate that our natural satellite is receding from us at about an inch and a half every year.

THE EFFECT UPON ECLIPSE TIMES

As it moves further away from us, the Moon takes longer to complete an orbit. Looking backwards in time, perhaps to 500 B.C., it was closer to us and so its orbital period was less. If the recession rate given above, an inch and a half per year, has continued throughout the intervening 2,500 years, this would imply that back then the Moon was about one-sixth of a mile closer, and the synodic month lasted for almost two seconds less than it does now.

Without accurate clocks in ancient times, how could we check the correctness of these calculations, which are based upon backward extrapolations of modern ultra precise measurements? The answer comes from eclipses. The milliseconds-per-day slowing of the terrestrial rate of rotation and the seconds-per-month discrepancies produced by the receding Moon, add together and result in severe displacements of the ground tracks of solar eclipse totality.

All those seconds accumulate to produce an eclipse time four hours earlier in 500 B.C., as foreshadowed above. This displaces the track of totality about 60 degrees east in longitude. For example, an eclipse that would otherwise have been expected to have had a track crossing the Italian peninsula 2,500 years ago might actually have been seen in Pakistan and western India, making a search through Roman republican accounts from that era futile. The problem can be attacked in the opposite way, however. With some ancient record of an observed total eclipse, knowing *where* it was observed, and approximately *when*, it is possible to back compute the circumstances of all feasible eclipses and identify the one responsible. Because the tracks of totality are so narrow, knowing a blacked out Sun was observed in Athens, Babylon, or Beijing on a certain date enables us to determine the spin phase of the Earth in that epoch.

There is a problem, though. If the rate of deceleration of the Earth's rotation were uniform, then the corrections needed would be straightforward. But this is not the case. Since the last Ice Age terminated about 10,000 years ago many continental regions (such as the northern parts of Europe, Asia, and North America), which were overlain for eons by thick ice layers, have been rebounding gradually. That is, their ice burden compressed them, but now they are expanding again. Like the skater raising her arms, this causes the spin rate to fall. One does not expect the deceleration in the Earth's rotation rate to be constant over millennial time scales, then. The eclipse records found on Babylonian clay tablets and in medieval chronicles are allowing investigators to track these changes in our planet's dynamical behavior rather precisely.

Having mentioned Beijing (formerly Peking) above, let us look at a specific record from ancient China. A couple of millennia back the imperial capital was Chang'an (known as Xi'an or Sian nowadays). A chronicle for 181 B.C. records that a total solar eclipse was witnessed there, and we can identify its circumstances through back-computations of the relevant orbits in all respects except one: the spin phase of the Earth. If one assumes that the planet rotated at its present rate throughout the years since 181 B.C. then the ground track of the eclipse would have missed Chang'an by about 50 degrees of longitude (equivalent to 3 hours and 20 minutes of spin), as shown in Figure 6-2. But the eclipse track *did* intersect Chang'an, indicating how much the Earth's rotation has slowed over all those centuries.

Many of the Babylonian clay tablets containing records of ancient eclipses are now archived at the British Museum, in London. It is not a coincidence that much of the leading work on ancient eclipse interpretation has been by British astronomers. In particular Richard Stephenson of the University of Durham, aided by Leslie Morrison of the Royal Greenwich Observatory and others, has found vital evidence for how the Earth's spin rate has varied since about 700 B.C. Mesopotamian tablets and Chinese records, plus various Arab chronicles and European annals, have all

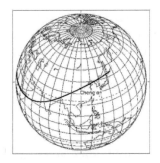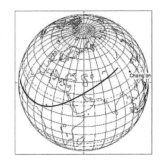

FIGURE 6-2. A total solar eclipse was observed from the ancient Chinese capital of Chang'an in 181 B.C. The eclipse ground track can be computed and plotted onto the globe such that it passes through Chang'an, as on the left, indicating the spin phase of the Earth in that era. If our planet had continued to rotate at the same rate as at present over all the intervening years, then the track would have missed Chang'an by 50 degrees of longitude (equivalent to 3 hours and 20 minutes of time), as on the right. Such eclipse records allow us to understand how the Earth's spin rate has slowed under tidal friction in recent millennia.

been trawled for their useful eclipse data. The results have applications in a number of areas of science other than just astronomy, for instance in developing our understanding of the long-term climatic vagaries of the Earth.

THE DISCREPANCY BETWEEN EARTH AND MOON MEASURES

We have seen above that tidal friction is causing the spin rate of the Earth to fall, and to compensate for that the Moon is receding from us. Direct measurements are made of two quite different things: the rotation rate of the planet (using historical eclipses for the long-term changes, and ultra precise radio astronomical techniques and so on for the short-term changes), and the distance of the Moon (through laser ranging). You might expect the results obtained from these distinct measurements to be in agreement, but that is not the case: there is a marked discrepancy between them. Why is this?

Let us go back to thinking in terms of the length-of-day (LoD) because it is the easiest parameter to understand. We have said that the LoD is increasing by around 1.7 milliseconds per century. So, as I write in the year 2001 the LoD is 1.7 milliseconds longer than it was in 1901, and in the year 2101 it will be (we anticipate) close to 3.4 milliseconds longer than it was back in 1901. The LoD is a quantity we can measure directly.

The rate at which the Moon is receding is also measured directly (from the retro-reflectors left on the lunar surface; see Figure 6-1), and that inch-and-a-half per year can be converted into the equivalent increase in the LoD that would result over a century. But when we do this, the answer is 2.3 milliseconds, and not the 1.7 milliseconds we might have expected. How does the discrepancy of 0.6 milliseconds arise? Some other process must be counteracting a part of the slowdown due to the tidal drag imposed by the Moon.

The answer to this puzzle seems to be related to the Ice Age cycle. At first sight it might appear that, as sketched earlier, the melting of the vast ice packs at latitudes beyond 40 degrees (which occurred around 10 millennia ago) would lead to a simple expansion of the rock and soil that had been compressed beneath them. Such an expansion would increase the average distance of these landmasses from the spin axis of the Earth, and so the spin rate would fall just as when the ice-skater spreads her arms. But it is not quite that straightforward.

It seems that since the ice burden melted (one can hardly call it polar ice because it covered about a quarter of the globe) the shape of the Earth *as a whole* has been changing due to the migration of the liquid water so released. Rotating objects are not spherical, but oblate; that is, they are flattened slightly, the distance pole to pole through the middle being less than that measured crosswise in the equatorial plane. Since the termination of the last glacial period it appears that the Earth has become a little less flattened (that is, it has become closer to spherical), with oceanic water moving away from the tropics and towards the poles. This means that the water involved is nearer to our spin axis, and so possesses less angular momentum. Overall, this boosts the planet's spin rate very slightly. The enhancement caused by shape change is equivalent to 0.6 milliseconds (per day per century). Subtracting that from the tidal drag imposed by the Moon and Sun, the value of 2.3 milliseconds, the overall change measured directly is 1.7 milliseconds.

I have just slipped something else in there. I mentioned the *Sun* imposing a tidal drag, as it surely does. Although the Moon is the major cause of the tides as such, it is the solar influence that produces the difference between the heights of spring and neap tides. Because of the Sun's effect, the angular momentum of the Earth–Moon system is *not* conserved precisely: the system is not completely isolated. This, though, is a minor complication. In fact, you can probably imagine what is happening here. The tidal friction due to the Sun produces a change that must be taken up by the orbital angular momentum of the *Earth*, and in consequence the mean Earth–Sun separation increases a little. But the change involved is minute, compared to what is happening in the Earth–Moon system.

THE PHYSICAL SIZE OF THE MOON

In all of our discussions of eclipses so far we have tacitly assumed that the intrinsic physical sizes of the Sun and the Moon are not changing appreciably.

The Moon is a rocky body. When it was young and hot it was slightly larger because by heating things you generally make them expand. That is why a glass jar may crack if you pour in a boiling liquid: the rapidly heating interior tries to expand against the cool exterior, breaking it asunder. Imagine you are making jam and have gotten to the point where you pour the steaming liquor of fruit, sugar, and pectin into the jars. You should be sure that you immerse those jars first in boiling water not only to sterilize but also to heat them, thus avoiding breakage due to temperature differentials.

Although the Moon may have been a little larger when young, it has long since completed its cooling and reached its equilibrium dimensions. (The puckering of its surface during this cooling and contracting from an initially molten state is thought to explain some of its peculiar surface features like cracks and rills—similar to the wrinkling of a prune as it dries out.) From the perspective of eclipse calculations, the physical size of the Moon may be taken to be unchanging. Over the time scale of human history, even the gradual increase in its mean distance from the Earth is not a significant effect compared to the monthly in-and-out movement varying its angular size.

THE PHYSICAL SIZE OF THE SUN

Turning our attention to the Sun, this is a gaseous body so we may expect it to expand and contract. For example, changes in the rate of energy generation (through nuclear fusion in its core) will cause its diameter to vary. The Sun is almost 5 billion years old, and over that time span we know that its energy output has not been constant. After the next 5 billion years, when the hydrogen fuel within it starts to be exhausted, astrophysicists expect the Sun to expand to become a red giant star, with a radius perhaps as large as the orbit of Jupiter, more than a thousand times its present size. After that, with little internal energy production to support it, the Sun will shrink again and attain a dimension rather less than at present, becoming a white dwarf.

Astronomers see these processes occurring in other stars and witness outbursts and oscillations in stellar sizes on all sorts of time scales. Some alter quickly, within days or weeks, but most stars have shown no significant alteration over the decades in which measurements of their brightness have been possible. Although the solar output is reasonably constant in the short-term (which is just as well, otherwise we might get fried), we should be prepared at least to entertain the notion that over centuries or millennia the Sun might grow or shrink. Such variations would of course affect the occurrences of eclipses, and their characteristics. In Chapter 7 we turn our attention to this matter.

7

Eclipses and the Size of the Sun

Observe due measure, for right timing is in all things the most important factor.

Hesiod, a Greek poet of the eighth century B.C.

The name of Edmond Halley has already appeared several times, in connection with the eponymous comet, his rediscovery and titling of the saros cycle of eclipses, and his suggestion that the salt of the sea could tell the age of the Earth. Now we are going to renew our acquaintance with him.

Having brought up his name, I should note that both parts of it have provoked modern dispute. Halley himself used two spellings for his given name: Edmond and Edmund. Whichever one might use is a matter of choice. Regarding his surname, the arguments have centered upon its pronunciation: "Hal-ee," "Haw-lee," or "Hay-lee"? The average person tends to go with the final version (mainly through familiarity with Bill Haley and the Comets, of "Rock Around the Clock" fame in the 1950s). However, the presence of the double "l" in Edmond Halley's patronymic indicates that one of those initial two pronunciations is actually more likely to be correct. The first is that most favored among astronomers. (I won't confuse the matter further by worrying over whether the second syllable should be "lay" or "lie" rather than "lee.")

HALLEY AND ANCIENT ECLIPSES

However we spell or say his name, Halley's interest in eclipses provides a bridge between the subjects of Chapters 6 and 7. In the previous chapter we saw that ancient eclipse records have allowed scholars to investigate how Earth's rotation rate has slowed over the past few millennia, with various astronomical and geophysical ramifications. More than three centuries ago Halley was interested in this apparent slowdown—he was the first to notice it—but from a rather different perspective.

In his era, appointments to university positions in Britain were heavily influenced by religious considerations. Various factors counted against Halley when he was an applicant in 1691 for the Savilian astronomy professorship at Oxford University. He even held the heretical view that comets (such as that bearing his name) could smash randomly into the Earth, causing great devastation. This did not fit in well with ecclesiastical views on divine providence. Mostly, though, his opponents were disquieted by his notion that the world might be older than the biblical chronology would indicate.

Learning from his failed application, Halley gained the religious-bias initiative in the following years through his study of ancient eclipses. In October 1693 he read a paper to the Royal Society ". . . concerning a Demonstration of the Contraction of the year, and promising to make out thereby the necessity of the world coming to an end, and consequently that it must have had a beginning, which hitherto has not been evinced from any thing, that has been observed in Nature." What Halley showed was that the times of eclipses spread over millennia could only be explained if the number of days within a year were reducing. This must indeed be the case, because the absolute duration of the year stays constant, but the days are lengthening, as we saw in the preceding chapter.

Halley's interpretation of this apparent elongation of the year, based on Christian dogma, was that the age of the world must be finite. The universe, he said, must have been a divine creation *ex nihilo* a handful of millennia before. The academic selection panel—the members were not only from within the University of Oxford, the Archbishop of Canterbury for example being among them—regarded this most favorably. They looked upon religious correctitude as being of the utmost importance, and Halley's careful demeanor during the 1690s had the end result that he was successful in obtaining appointment to the Savilian Chair of Geometry in 1704.

Halley was skilled at computing the past tracks of totality over foreign lands after his earlier work. Looking forward, he recognized that in 1715 a total solar eclipse would sweep across southern England and Wales, the first time that London had been so-visited since 1140 (and 878 before that). He turned his hand and mind to computing its precise course, and organizing observations.

A detailed predictive map that Halley prepared is shown in Figure 7-1. A pamphlet that was widely disseminated at the time, showing such a map, was entitled The Black Day or a prospect of Doomsday exemplified in the great and terrible eclipse which will happen on the 22nd of April 1715. If the simple information that

FIGURE 7-1. The ground track over England and Wales of the solar eclipse of 1715, as computed ahead of time by Edmond Halley. In reality the track was slightly wider. This was by just a few miles at the northern extreme, but with a southeasterly displacement of about 20 miles for the southern boundary (compare this pre-eclipse prediction with the post-eclipse map, also drawn by Halley, as shown in Figure 7-2).

an eclipse was to occur didn't rustle up public interest, that pamphlet was sure to do so.

THE CAPITAL ECLIPSE OF 1715

A total solar eclipse has not crossed England's capital city since that day in 1715. Nor will London see such an event again for some time to come. Three centuries ago, the eclipse ran from about eight until ten in the morning, with totality lasting for a few minutes at around ten past nine. At least, that was the time in London. Not only did the shadow reach other locations at different absolute instants of time, but also in those days there was no standard time in Britain, each town keeping its own clock time according to the Sun's position, making the nationwide comparison of observations difficult. Regarding the *date* of the eclipse, we will come to that at the close of this chapter.

Nowadays any eclipse is gazetted well in advance, so that amateur and professional observers alike are well prepared, but that was not the case in Halley's era. He wrote to a wide variety of potential observers. From the rectors of village churches and the like he received a flood of useful information, allowing him to determine the path of totality with admirable accuracy. Not only that, but the comparative timings for the duration of the eclipse were very useful check readings, as these would be longest near the central line, dropping to zero at the edges of the path, and the duration would also vary *along* the track due to the Earth's curvature.

The amateur observers did well then—but what about the professionals? In Cambridge, the Plumian Professor of Mathematics, Roger Cotes, tried to time the eclipse but was distracted by what he termed "too great Company," and so he did not obtain the necessary data. Halley himself was in London for the eclipse, gathered with various other fellows of the Royal Society. That is just as well, because Oxford was clouded out. Under Halley's guidance, his group obtained useful timings.

If central London were clear, as it was, then one would anticipate that the astronomers at the Greenwich Observatory must also have made detailed observations. They may well have done so, but in a spirit of fine scientific collaboration the Astronomer Royal, John Flamsteed, refused to allow Halley direct access to the Greenwich data, which were never published. Halley succeeded Flamsteed as Astronomer Royal in 1720, and it is surprising that he did not himself dig out the 1715 eclipse observations thereafter, although to be fair he was always busy with new scientific tasks.

Be that as it may, what Halley really needed was not lots of observations from just one place, but rather information from a wide geographical scatter. That way he would be able to determine the width of the ground track. If some curate standing in his churchyard saw a brief instant of totality, and yet the verger sent to the crossroads in the village a mile to the east did not, then Halley would know that the edge of the shadow had passed between them. Thus the precise positions of the observers, plotted onto a map, were important.

This is just what Halley got. Some dozens of reports were supplied by correspondents scattered over England and Wales, enabling him to determine the northern and southern extremities of the track to within a mile or so. For example Halley was soon writing:"From these observations we may conclude that this Limit came upon the coast of England, about the middle between Newhaven and Brighthelmston [Brighton] in Sussex." Similarly he found that in Wales the northern limit ". . . entred on Pembrokeshire about the middle of St Brides Bay."

Comparing these points with Halley's pre-eclipse prediction (Figure 7-1) we see that he was inaccurate, by only 3 miles for the northerly limit, but by 20 miles for the southerly. The ground track that Halley determined from the observations is shown in Figure 7-2; it was about 183 miles wide, 23 more than his prior estimate.

That might initially seem peculiar. One could understand the track being uniformly displaced in one direction or another due to slight timing errors, but how could its *width* be wrong? The answer lies with the lack of precise evaluations of astronomical distances in that era. Later in this book we will discuss how James Cook was sent to the South Pacific in 1769 specifically to watch the transit of Venus across the face of the Sun, as part of an attempt to measure more accurately the mean solar distance from the Earth. Halley was one of those who invented the technique employed in that episode. Back in 1715, Halley could not be sure of the distances and sizes of either the Sun or the Moon, and in consequence he substantially underestimated the width of the eclipse track.

THE IMPORTANCE OF KNOWING THE SUN'S SIZE

No area of science has made recourse to historical information more often than astronomy, and Halley's 1715 eclipse compendium is a wonderful example, as we shall see. First, though, I must sketch in a little of the background.

Halley was only 19 years old when he first made observations of sunspots, publishing the results in his second scientific paper.

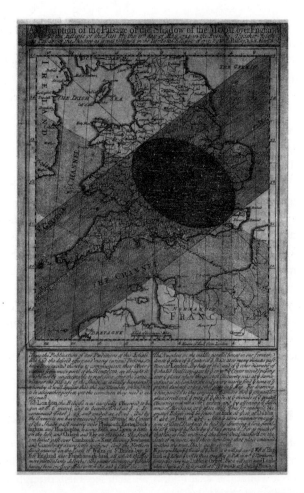

FIGURE 7-2. The actual total eclipse track observed in 1715, as assembled by Edmond Halley from eyewitness reports, along with his predicted path for the 1724 eclipse (the track slanting downward from upper left to lower right).

That was in 1676. Using the naked eye, the ancient Chinese had observed large sunspots many centuries before that, when dust storms blew in from central Asia, blanketing parts of northern China. Such solar blemishes had similarly been noticed from Europe, but it was only when telescopes appeared in the seventeenth century that continuous monitoring of these dark markings on the Sun's surface was feasible.

Using a telescope an image of the Sun can be projected onto a screen (as in Figure 1-13). By following the movement of specific sunspots from day to day Halley and his contemporaries determined that, near its equator, the Sun takes about 25 days to spin. Not being a solid body, it does not rotate rigidly, but different speeds are apparent depending on the latitude, such that nearer its poles the Sun takes closer to 35 days to turn once.

Sunspot numbers have routinely been kept through to the present from Galileo's time, a hundred years earlier than the eclipse in question, and it was studies of these numbers that revealed the apparent 11-year periodicity in solar activity. There is evidence that the overall climate of the Earth follows the same cycle. In the early decades of the twentieth century another British astronomer, Edward Maunder, noted there had been a deficit of sunspots during the latter half of the seventeenth century; this is now known as the "Maunder Minimum." This coincides with a pronounced cooling of the climate known as the "Little Ice Age." The River Thames froze over, for example, and fairs were held on London's ice-covered waterway. This correlation may just have been a coincidence, but it seems to warrant more than merely a suspicion that the two phenomena are related.

This makes one wonder how else the Sun might be varying in its properties, sunspot numbers being just one diagnostic, and how the Earth's climate might alter in response to any such change. If the Sun had expanded, say, then one might also expect it to cool a little, and so not emit so much energy in the form of sunlight. There would then be a concomitant drop in the mean temperature of our planet. Certainly, astronomers observe other stars pulsating in and out, their power output varying radically, and because our climate balances on a knife edge a fairly slight alteration in the Sun's power could have major repercussions.

Studies of stellar evolution indicate that since it "switched on" over 4.5 billion years ago, the energy output of the Sun has increased by about 30 percent. In fact, this understanding is so well established that the jargon phrase "Early Faint Sun Paradox" is bandied about within circles of scientists interested in the evolution of the terrestrial environment, and in particular those studying how life developed on our planet. The point here is this: If the Sun were initially so much fainter, as is believed to be the case, then the Earth would have been a frigid world. Under such circumstances, how did even the simple mono-cellular slime, which was the sole occupant of the planet between about 3,800 and 570 million years ago, manage to evolve and survive?

This increase in solar output has not terminated. In our earlier description of solar evolution it was noted that the Sun is expected to continue to behave in a similar fashion to the present for another five billion years or so. Over that time, though, its power output is expected to double. If that increase were steady and uniform then over five thousand years (a suitable time scale for human civilization) the solar energy reaching the Earth might increase by one or two parts in a million. Such changes are dwarfed by other natural variations, like the way in which the Earth's orbit evolves and the orientation of its spin axis shifts. But what if the

Sun's output oscillates significantly or alters abruptly on timescales of only decades or centuries? Astronomers certainly see other stars acting in this way.

Such modern concerns as the anthropomorphic "greenhouse effect" would obviously be affected by changes in the solar output, so we'd better be sure we know how the Sun behaves over extended periods. Perhaps historical measures can assist. At last, then, we come to the significance of Halley's compilation of reports of the 1715 eclipse.

A MODERN REANALYSIS OF HALLEY'S RESULTS

In the late seventeenth and early eighteenth centuries astronomers at the Paris Observatory made micrometer measurements of the apparent solar diameter, and in the 1980s French scientists compared these with modern values. They concluded that, if the old measures were correct, then three centuries ago the Sun was about one part in 480 larger than it is now. That is an appreciable fraction from the perspective of the possible climatic effect.

The problem is this: How accurate were those observations made 300 years ago, barely a century after the telescope was first used to peruse the Sun? If they were good, then the Sun must be shrinking, which might cause it to heat up, flooding the Earth with an increased flux of sunlight, thus adding to the greenhouse effect. Alternatively, if those early measurements of the Sun's size made from Paris were imprecise, then we might be able to discount such a possibility.

What was needed to solve this question was some alternative determination of the solar diameter from a few centuries back, but of greater precision. The precision attainable from direct measurements, however, was limited by the available technology, whether the telescopes used were French, Italian, or British. Some analogue determination of greater accuracy was required: an *indirect* measure.

Halley's detailed account of the reports he received of the eclipse from far-flung parts of Britain provides just such an analogue, as was realized in 1988 by Leslie Morrison and Richard Stephenson (whose work on old eclipses has already been mentioned), along with their colleague John Parkinson. If the Sun had been larger by almost 0.2 percent in 1715, then the limits of totality would have been about 6.5 miles narrower, just over 3 miles at both northern and southern edges. Coupled with our modern knowledge of the distances and orbits of the Sun and Moon, the actual observations of totality (or lack of it), which Halley preserved verbatim, enabled this team to determine the edges of the track to within a few hundred yards. As the Moon has not changed size, the small residual uncertainty implies that the Sun has not shrunk by as much as one part in 20,000 since 1715.

Halley's remarkable records of the observations of the eclipse of 1715 remain not only an exemplar of a great eighteenth-century scientist at work, but also the best evidence we have that the Sun has not greatly altered in size over the past several centuries.

ANOTHER ECLIPSE OVER THE BRITISH ISLES IN 1724

Much of southern England had waited from 1140 until 1715 for an opportunity to witness a total solar eclipse, about twice the average waiting time for random locations in the Northern Hemisphere. The inhabitants did not need to wait much longer, though, for the next show. On May 11, 1724 the shadow of totality swept diagonally across the southern halves of Ireland and Wales before proceeding in a southeasterly direction across most of England below Birmingham, as in Figure 7-2. Only Cornwall and Kent (the southwestern and southeastern tips of England) missed out beyond the southern limit, the northern edge of totality just missing Oxford and London, which still awaits a repeat of the Capital Eclipse of 1715.

Before leaving these eighteenth-century eclipses, a peculiarity should be mentioned. When the path of the 1724 shadow left Britain and crossed the English Channel into France, the date suddenly jumped from May 11 to May 22. Similarly the great eclipse of April 22, 1715, was seen on that date in Britain, but elsewhere it was May 3rd already.

What am I getting at? My point is that in Halley's day Britain was still using the Julian calendar, the 11-day jump necessary to fall in line with the Gregorian calendar not being made until 1752. The same is true for the American Colonies ruled by Britain in those days. Although history books may tell you that George Washington was born on February 22, 1732, in fact he was born on February 11, 1731 (because New Year for the British was not until March 25). Great Britain and Scandinavia were the last places in Western Europe to reform their domestic calendars, whereas in Eastern Europe it took some countries until the 1920s to make the change, by which time the discrepancy had grown to 13 days.

8

The American Eclipses of 1780 and 1806

June 16, 1806: Pleasant morning—total Eclipse of the Sun &
the stars twinkled at noonday. Wonderful are the changes of
nature but more astonishing the wonders of redeeming love.

Entry in the diary of Mary Avery White (Boylston, Massachusetts)

lsewhere in this book we have seen that superstitious beliefs concerning eclipses have enabled certain nations or armies to gain an advantage over their opponents. Let me add another two examples to the litany.

Ever since Byzantium had been adopted early in the fourth century by Constantine the Great as the capital of the Eastern Roman Empire, that city (renamed Constantinople) had been in Christian hands, despite internecine squabbling. The residents largely believed, however, in an ancient prophecy that said the city was safe from its enemies during the waxing phase of the Moon. In May 1453, while defending their walls against the marauding Ottomans, the Byzantines were horrified to see that the rising moon was in eclipse. Their morale was broken and a week later the city fell into the hands of the Turks, who have held it ever since. It was the capital of the Ottoman Empire, and then Turkey, until 1923 (when the capital was shifted eastward to Ankara), and then renamed again in 1930 as Istanbul.

That was an instance of an eclipse aiding the Ottomans. During the First World War, a counterexample occurred. On July 6, 1917, Lawrence of Arabia and his Bedouin troops overran the ancient city of Aqaba, located at the northern tip of the Red Sea. One of their advantages was that, having crossed the Sinai by camel, they attacked from the inland side, whereas the Turkish armaments were pointed out to sea to repel a maritime assault. Their other advantage came from the fact that as they approached from the north, on the evening of July 4 there had been a total lunar eclipse. This preoccupied the defenders with banging together pots and pans and otherwise making loud noises in order to scare off the shadow that was darkening the Moon.

Those were both lunar eclipses. In this chapter we will be considering the significance of solar eclipses in the early history of the United States. In Chapter 1 we learned that the ancient Chinese would shoot arrows into the air in order to scare away the dragon they imagined to be devouring the Sun during an eclipse. Similarly, the Native Americans of the Chippewa/Ojibwa tribes thought that the Sun's flames were being extinguished, and so during an eclipse they would launch skywards burning arrows in order to replenish it. We will see below that the total solar eclipse of 1806 was one of the pivotal junctures in the Indian wars provoked by the spread of the white man westwards through Ohio and Indiana.

THE ADVENT OF ECLIPSE TRACK PREDICTION

Edmond Halley was the first person able to predict with reasonable accuracy the tracks of total solar eclipses, such as those crossing the British Isles in 1715 and 1724. This was an important development, not just for science. It meant that, in principle, it was feasible for cultures with modern scientific attainments (and thus the ability to read an astronomical almanac) to scare the wits out of less-learned peoples.

The reality is that this ability was only really employed by the authors of novels, such as H. Rider Haggard, Mark Twain, and Hergé, whose invention of solar eclipses at critical points in their stories were considered earlier. For potential conquerors or colonists the problem, as such, was that total solar eclipses are so infrequent that it is most unlikely that a track will pass through any region of interest where they are trying to unseat the natives. Lunar eclipses—like that advantageously interpreted by Christopher Columbus in 1504 (see Figure 8-1)—may more easily be used because they can be seen over a much wider area.

Given these facts, it seems ironic that the two total solar eclipses to cross the fledgling United States in the decades following independence—in 1780 and in 1806—each had results that are the converse of what one might have expected. Their stories contradict the impression that the superior science delivered by Halley and his successors must surely have lead to advantages for the purveyors of precise astronomical knowledge. The first eclipse led to a major embarrassment for the new American astronomy. And it was the indigents, rather than the incomers, who exploited the second eclipse to gain the upper hand in a festering dispute.

THE ECLIPSE OF 1780

Soon after its foundation in the seventeenth century, Harvard University instigated the study of physics. In 1726 a benefaction from

FIGURE 8-1. This drawing of Christopher Columbus using the lunar eclipse of February 29, 1504, which appeared in Washington Irving's *Life and Voyages of Christopher Columbus* in 1892, gives a rather different impression to that shown in Figure 3-5, which is from a few decades earlier. This depiction shows Columbus in a more benevolent light, garbed in fine clothes and with the Jamaicans at his feet, whereas the other picture has him armed with a sword and closely accompanied by armed guards.

an Englishman, Thomas Hollis, led to the endowment of a professorial chair of "Mathematicks and Experimental Philosophy." It continues to this day.

The second occupant of that chair was John Winthrop, who was appointed in 1738 at the age of 24 and held the position until his death at 65. Winthrop is often regarded as the first true "American astronomer," and he made observations of many celestial phenomena. In particular he was involved with timing the transits of Venus, a subject we discuss in Chapter 13. It seems though that

Winthrop's attainments in physics were not so wonderful, and it is not clear from the notes he left that he even understood Newton's laws of motion, one of the most rudimentary facets of the science. Winthrop's demise led to the appointment of Samuel Williams, just in time to start planning for the solar eclipse anticipated for October 27, 1780. The track of totality was expected to pass over much of Maine and parts of maritime Canada.

There was an obvious problem. This was the time of the Revolutionary War, and the track lay within enemy (i.e., British) territory. Undeterred, Williams made his calculations, studied his maps, and chose the western part of Penobscot Bay in Maine as a suitable observation point. This choice was based largely on the need to bring in a large sailing ship carrying the heavy equipment required for the eclipse observations: the telescopes, clocks, and so on. This decided, Williams prevailed upon John Hancock, the first signatory to the Declaration of Independence and in 1780 the Speaker of the Continental Congress, to write to the commander of the British forces. "Though we are political enemies, yet with regard to Science it is presumable we shall not dissent from the practice of all civilized people in promoting it," wrote Hancock. After such sweet-talking, safe passage was granted to the party.

On October 9 a group of four faculty members and six students set off up the coast in a boat supplied by the Commonwealth of Massachusetts. On arrival in Penobscot Bay they set up their equipment, calibrated their clocks with other astronomical observations, and confidently awaited the eclipse. This duly arrived, starting in the middle of the morning and reaching a peak shortly before noon. There was just one problem: They did not witness totality. The visible fraction of the solar disk shrank to a sliver, but it did not disappear. Williams's calculation of the eclipse track was wrong. It turned out that they should have been positioned at least 30 miles further north.

The embarrassment over the failure of this official expedition was made even more acute by the fact that on the British side Dr. John Clarke, a Harvard graduate, had successfully witnessed the eclipse from Prince Edward Island. Accompanied by Thomas Wright, the local surveyor general, Clarke used a small telescope to make observations, and then sent the results to the American Academy of Arts and Sciences.

What went wrong with the expedition led by Williams has long been a mystery. In an attempt to throw light on the matter, on the bicentenary of the eclipse in 1980, a new expedition of Harvard staff and students returned to Penobscot Bay, taking with them maps and instruments identical to those used by Williams and his group. All worked precisely as they should, although there have been questions raised as to whether the times of the eclipse contacts reported by Williams, and the extent of the arc of the Sun seen as remaining uncovered, were consistent with the stipulated observatory site.

Explanations for the blunder fall into three categories. The first is that Williams simply made a numerical error in his sums. The second is that the map used was inaccurate, showing the wrong latitude for the bay. The third category is connected with errors in the astronomical tables from Europe used by Williams. Although there seems to be inadequate documentary material left in order to know whose fault this farce was, it has been usual to lay the blame at the door of Samuel Williams. Both during this first American eclipse expedition, and in later life, he did a number of things that might be regarded as imprudent. But there is not the evidence necessary to be sure of his guilt here.

There is someone else who was greatly unhappy at turns of events on October 27, 1780, in this case for an entirely different reason than the failed eclipse expedition. This story involves another of the signers of the Declaration of Independence, Francis Hopkinson. Despite a widespread belief that Betsy Ross designed the Stars and Stripes, in fact it was Hopkinson who was responsible some years earlier. He invented not only the basic design for the flag of the United States, but also various other insignia and seals. With some justification he felt it was his due that this should be recognized and some nominal amount paid to him. On the same date as the eclipse the Treasury Board presented its report to Congress and recommended that because Hopkinson was already in the government's employ he should receive no further payment. In consequence he resigned in disgust. After all, he had only asked for a "quarter cask of the public wine" as a token of gratitude.

BAILY'S BEADS SEEN IN 1780

One may or may not consider another outcome of the eclipse expedition to Maine in 1780 to be an embarrassment. In the opening chapter we met Francis Baily, the British astronomer who in 1836 gave a description of the luminous phenomenon seen just as totality begins and ends, universally known as "Baily's beads." It so happened that Samuel Williams noticed these beads of light during the 1780 eclipse. If his account had been better known, then people would nowadays talk about spotting "Williams's beads," and he would not have been relegated to the fringes of science history. Those few tens of miles made all the difference.

In fact there is more to it than that. Paradoxically, the mistaken location of the Harvard expedition made it more likely that they would witness the beads than if they had been right in the middle of the eclipse track. Ever since Halley made eclipse track prediction possible, astronomers have glibly assumed that the best place to be is right on the central line, but in recent years it has slowly dawned on eclipse watchers that this is not the case at all.

Imagine that a particular eclipse track is precisely 100 miles wide, and the maximum eclipse duration is to be 3 minutes. There is 50 miles from the center to the edge and beyond that edge there will be no totality. If you are positioned 10 miles from the central line, the duration of totality falls by only a few percent, and so there is no need to worry about one's precise position. If, though, you are located 47 or 48 miles from the central line (2 or 3 miles from the edge) then totality will last for only one minute, rather than three.

That's an assessment of the *quantity* of totality; a quite different thing is the *quality*. It happens that total solar eclipses are rather more spectacular if they are viewed from near the periphery of the path. Because they depend upon the light just reaching your eyes around the crinkly edge of the Moon, both the diamond ring effect and Baily's beads may last for about ten times longer if you are near the fringe of the track. Indeed, the beads often seem to run quickly around one edge of the Moon in that position. The elusive *shadow bands* discussed in Chapter 15 may last for up to five times longer and be easier to see. But more important still is your chance to see the chromosphere, the most colorful feature of an eclipse.

For decades astronomers labored to get good photographs of the chromosphere, and in particular its spectrum. Close to the central line, the opportunity to obtain such observations lasts for only a few seconds and because of that it is often termed the "flash spectrum" (see Figure 5-7). If only they had positioned themselves well away from the central line, the astronomers involved would have stood a far better chance. The reason is this: If you are located near the very edge of the eclipse track the apparent disks of the Sun and the Moon glide along the same tangential line. In consequence the bright red chromosphere may be visible for up to a minute and a half, allowing ample time for its spectrum to be captured. Similarly, giant prominences can be seen for longer, jutting up above the solar surface.

Although they experienced no actual totality, Samuel Williams's party was close enough to the edge of the track that only a very thin layer of the Sun was visible. This was slender enough that it appeared to become broken up into separate patches by the mountainous limb of the Moon, and so they gave an account of the phenomenon that later became known as Baily's beads. They were aided by the fact that they were viewing by telescope: it is easier (but dangerous, unless you know precisely what you are doing) to see the beads with a telescope, rather than the naked eye.

It is worth noting in passing that Francis Baily could perhaps have been aware of the discussion of the eclipse by Williams. In 1796 a youthful Baily spent some time in America, traveling west from the east coast and then down the Mississippi by boat to New Orleans. From there he walked much of the two thousand miles back to New York. His published account of that trip is a classic of the era. In those days, however, his knowledge of astronomy was still modest. It was only after his return to England, and his amassing of a small fortune in business, that he was able to devote his time to the study of celestial phenomena and assist in the founding of the Royal Astronomical Society.

THE GREAT DARK DAY OF 1780

Every year there are several eclipses and 1780 was no exception. Indeed a partial lunar eclipse could be seen from New England, early in the morning on May 18, although only at moonset. The following day was most peculiar, though. It has gone down in history as the Great Dark Day. No one really knows what caused the darkness experienced then, throughout the states in the northeast.

The basic information is clear enough. At about 10 in the morning on May 19 the sky started to dim and by 11 there was darkness all around. It seems that the whole of New England was affected, an area "at least 650 miles in extent" according to contemporary reports. The Sun was blanked to such an extent that it was impossible to read a newspaper. In the context of this book, it is worthwhile to note that the responses of plants and animals were the same as during a solar eclipse. Cows ambled back to their sheds, fowl went to their roosts, bees returned to their hives, other insects went quiet, and flowers closed their petals.

It was not only the Sun that was affected. As there had been a lunar eclipse the day before, obviously the Moon was just past being full. When it rose on the evening of May 19, it was dimmed too. This gloomy state of affairs continued until two o'clock on the morning of May 20, and by four all had returned to normal, except that people were mightily upset. This is what the *Boston Independent Chronicle* related: "During the whole time a sickly, melancholy gloom overcast the face of Nature. Nor was the darkness of the night less uncommon and terrifying than that of the day; notwithstanding there was almost a full moon, no object was discernible. . . . This unusual phenomenon excited the fears and apprehensions of many people. Some considered it as a portentous omen of the wrath of Heaven in vengeance denounced against the land, others as the immediate harbinger of the last day, when 'the sun shall be darkened, and the moon shall not give her light'." The closing quote there, from the Bible, shows the way many people regarded the dark day in 1780. Indeed the event is still cited by some with strong religious beliefs as having been a sign of the Second Coming.

Five months later the sky went dark again around noontime in New England, but the reason for that is well understood. The eclipse on October 27 was seen by many. For a few, though—the party of ten from Harvard camped in Penobscot Bay and surrounded by an opposing army—the sky did not get as dark as they had hoped or earnestly expected.

INSPIRED BY ECLIPSES

One of the great figures of American science in the early nineteenth century was Nathaniel Bowditch. Born in Salem, Massachusetts, in 1773, this self-educated mathematician and astronomer made many seminal contributions to the physical sciences, but he is best remembered for his book entitled *The New American Practical Navigator*. This volume has gone through more than 60 editions, updated long after his death in 1838 so as to include developments such as the use of radio and electronics, but it is still commonly known by the simple sobriquet "Bowditch." At the age of seven Bowditch may well have witnessed a partial eclipse from his home when totality swept its path a little to the north in 1780. In 1806, though, he certainly observed a total eclipse from his garden in Salem.

A youngster who did see the eclipse in 1806 wrote about it

six decades later, in the *Boston Globe*, with a familiar recounting of the effect of the eclipse upon animals: "Then a boy of five and a half years old, [I recall] seeing the eclipse through a piece of smoked glass, as I was held up in my father's arms in the yard of our home on Washington Street, Boston. Our hens made for the barn, and the doves flew to their cote; the cows, then by custom pastured on Boston Common, sought the Park Street outlet, supposing milking time to be at hand, but the returning light disabused them of their error, and they again took to grass."

In this later age the date June 16 has attained a specific meaning for many people. It is often termed "Bloomsday" because that was the date in the year 1904 on which all the action in James Joyce's novel *Ulysses* takes place, following the life of the central character, Leopold Bloom of Dublin. Go to an Irish gathering anywhere in the world on that date and you will find public readings from *Ulysses* underway.

Another budding writer gained some inspiration from the eclipse of June 16, 1806, James Fenimore Cooper. The town of Cooperstown, which houses the Baseball Hall of Fame, gets its name from Cooper's family. Located on the shore of Lake Otsego in upstate New York, Cooper was there, about two hundred miles west of Bowditch in Salem, when the eclipse passed by. At the age of 16 young James could and should have been away at college, but Yale had expelled him for fighting, and he had been packed off home with his tail between his legs. The eclipse certainly taught him something. Later he would write: "Never have I beheld any spectacle which . . . so forcibly taught the lesson of humility to man as a total eclipse of the sun." Cooper's experience was heightened by a melancholy circumstance. A local schoolteacher had been convicted of murdering one of his pupils, and after a year of confinement in a windowless cell awaiting his fate, the prisoner was allowed out to see the eclipse. Cooper was taken to see this unfortunate man, who was weighed down with chains and looking haggard and pale. The combined experience was almost more than Cooper could bear, and eclipse allusions appear in several of his later novels.

THE CORONA UNDERSTOOD

The quotation from Edmond Halley that began Chapter 5 indicates that, in his day, the corona seen during an eclipse was assumed to be a lunar—rather than a solar—phenomenon. The early modern astronomers Clavius and Kepler thought that the corona was simply the back-illuminated atmosphere of the Moon, although one could raise any number of objections to that interpretation.

The fact that the corona is actually the extended but tenuous atmosphere of the Sun became clear from observations made of the eclipse of June 16, 1806. The Spanish astronomer José Joaquin de Ferrer traveled to Kinderhook, just south of Albany, New York, to observe. You may have noticed that "corona" appears to be a Spanish word, and this is why: it was Ferrer who coined the term to describe the crown he saw circling the Sun during totality that day. The pivotal scientific contribution he made stemmed from his measurement of the extent of the corona. Ferrer showed that if it were a lunar atmosphere then it must be 50 times larger than that of the Earth, a notion that made no sense. In consequence he opined that the corona was attached to the Sun. The final proof of this did not come until much later, when the first eclipse photographs were obtained in the 1840s and 1850s, but it was Ferrer who first recognized the origin of the corona.

That is not to say that Ferrer neglected the Moon. Especially in view of our discussion about Baily's beads, it is worth mentioning that the Spaniard also noticed that the irregularities of the lunar surface could plainly be discerned around its periphery. Those, of course, are the cause of the beads of light often seen as totality begins and ends.

The eclipse track proceeded eastwards, then, from Cooperstown to Kinderhook to Salem, plus all places in between. It is sometimes misstated that this was the last total solar eclipse visible from New York City prior to that in 1925, which we will discuss in Chapter 10. However, that is incorrect. Although the eclipse was total in a band stretching across upstate New York, the people in the city never saw less than one-sixtieth of the solar disk exposed, and you have to go several centuries further back—to before the foundation of the city—to find the preceding event.

But what of the track of the eclipse of 1806 further west, in the region of the Great Lakes?

THE ECLIPSE ON THE WESTERN FRONTIER

Nathaniel Bowditch wrote his book about navigation largely as a result of finding over 8,000 errors in an earlier manual, written by an Englishman, John Moore; hence the inclusion of the word "American" in the title of his book. Volumes such as that by Moore were in wide circulation, in order to enable explorers to find their way in uncharted territory, as well as for sailors to navigate at sea (so long as the mistakes did not lead them astray). All sorts of astronomical phenomena would be listed: things like the times and circumstances of eclipses, and occultations by the Moon of bright stars. These would be invaluable to the hardy souls pushing west over the Appalachians and beyond in the quest for new lands to settle.

Those lands were occupied, of course. American Indians had lived on the Great Plains for millennia before the white men arrived. Friction and strife were inevitable. This is not the place to detail the history of the wars and battles that occurred as the settlers usurped the ancient territories of the indigenes. We are interested here simply in the eclipse of 1806, and how it enters into that greater story.

Prior to it reaching New England as described above, having passed over much of upstate New York and the northern half of Pennsylvania, the eclipse track had enveloped the whole of Lake Erie. Using the states as later delineated, the north of Ohio was crossed, and before that the south of Michigan, and half of Indiana. The angled path of the track put the northernmost fringe near the present city of Gary, the southern near St. Louis, with Fort Wayne fairly close to the middle.

The people in these regions should not have been entirely ignorant of the forthcoming eclipse. As foreshadowed above, many settlers and explorers would buy an annual almanac, containing notes of what was to be expected in the forthcoming year. Although it is usually remembered for its humorous aphorisms, a half-century earlier Benjamin Franklin's *Poor Richard's Almanac* had been one of the biggest-selling books in the American colonies, and others had followed in its stead.

Perhaps as one went further west, the frontiersmen had other things to think about. But the eclipse in 1806—which might have been used as a tool to quell the natives as Columbus had done in Jamaica three centuries before—was actually used by the Indian people to provoke an uprising against the insurgent whites.

THE SHAWNEE PROPHET

Although they had earlier fought against the British and the French, after the Declaration of Independence in 1776 the Indian nations in the Old Northwest (the area to the east of the Mississippi and north of the Ohio River) tended to side with the British against the newly defined Americans. Perhaps the greatest native leader to have arisen since Europeans first arrived in North America was Tecumseh of the Shawnee. Born near Springfield, Ohio, in 1768, Tecumseh eventually died in 1813 at Battle of the Thames, near Detroit, fighting alongside the British with his Indian warriors.

Tecumseh was one of eight children. One of his brothers, ten years younger than he, was named Lauliwasikau, which meant "Loud Mouth." Apparently he was a noisy baby, and one of triplets at that. Unlike Tecumseh who was a renowned warrior and thinker—he read the Bible and also books on world history—Lauliwasikau was a dissolute character. As a child he had been blinded in one eye in a hunting accident, and he fell steadily into alcohol over-consumption as he grew older. In 1805 he drank himself into such a stupor that his family thought he was dead and began preparing a funeral pyre. To considerable surprise Lauliwasikau suddenly awoke, saying that the Great Spirit had shown him wonderful visions. He then foreswore liquor and all other appurtenances of the white man and declared that he was an instrument for the Indian people to lead their way forward. Tecumseh had long realized that for the American Indians to survive against the encroachment of the whites from the east it would be necessary for all the tribes to band together in a common purpose. His efforts in this regard had been stymied by inter-tribe

FIGURE 8-2. Tenskwatawa, the Shawnee Prophet, used foreknowledge of the eclipse of 1806 to stir up unrest among the American Indians of Ohio and Indiana.

rivalry. Now, though, Tecumseh saw in his brother the instrument to unite the Indians to resist the otherwise inevitable, gradual seizure of their lands.

Lauliwasikau's name was changed to become Tenskwatawa, meaning "He Who Opens the Door." He is more often recalled, though, as The Shawnee Prophet (Figure 8-2). He was a shaman, a charismatic religious leader whose influence quickly spread through Ohio and Indiana. He claimed he could cure all types of disease and provide divine protection in battle for his adherents, a popular notion that the people were ready to believe.

Tenskwatawa himself was influenced by the Millennial Church, which had originated in eighteenth-century England, but owing to persecution had migrated across the Atlantic in 1774 to seek a more tolerant horizon. The members are usually known as the Shakers, due to a ritual dance they perform involving a shaking motion of the body. Thus the Prophet was directly affected by an English connection, and he sought to build up anti-American feelings to the benefit of the morale of his own people. He was very clear in differentiating between the different peoples of European origin. In 1807 he said: "I am the Father of the English, of the French, of the Spaniards and of the Indians. . . . But the Americans I did not make. They are not my children but the children of the Evil Spirit. . . . They are very numerous but I hate them. They are unjust—they have taken away your lands which were not made for them."

THE ECLIPSE PREDICTED

The Prophet first came to the attention of the United States administration when he was involved in the burning of some proclaimed witches. The Delaware Indians, originally from the region now known as New Jersey and the state bearing their name, had become refugees through the grabbing of their lands. In consequence they had been driven westwards, into Ohio. They were not happy. Hearing about how the Prophet had condemned both the Americans and also other religious leaders, they invited him to come and help them purify themselves. Sure enough the Prophet identified several of their number as witches responsible for the ills that had befallen them and ordered their torture. One unfortunate woman was roasted for four days over a slow fire. The Indians accused of witchcraft tended to have one thing in common: they had taken up with at least some of the ways of the whites (for example wearing hats or drinking liquor). Christian converts among the tribes were a particular target. Tenskwatawa was weeding out those who would oppose his brother's campaign for a return of the Indians to their olden ways. He moved on to other villages, and other tribes, stirring up revolt as he did so.

This unrest in the Midwest was viewed askance from Washington. Thomas Jefferson opined that the Prophet was "more rogue than fool, if to be a rogue is not the greatest of all follies." Although Tenskwatawa was trying to persuade or force his fellow Indians to give up the ways of the white man, Jefferson did not view him as being a major threat: "I thought there was little danger of his making many proselytes from the habits and comfort they had learned from the whites, to the hardships and privations of savagism, and no great harm if he did. We let him go on, therefore, unmolested."

The governor of the Indiana Territory was William Henry Harrison, who in 1841 became the ninth President of the United States. He had a high opinion of Tecumseh, once writing to the Secretary of War that the Indian leader was a "bold, active, sensible man, daring in the extreme and capable of any undertaking." Now, though, Harrison was faced with Indian roguery that was reaching a fever pitch, under the guise of divine influence. It became obvious that something would need to be done.

Early in 1806 Harrison wrote an explicit, challenging letter to the Delawares, inviting that they demand the Prophet prove his exalted status. Using phrases from the Bible, Harrison suggested to them that they should "ask of him to cause the Sun to stand still, the Moon to alter its course, the rivers to cease to flow or the dead to rise from their graves. If he does these things, you may then believe he was sent from God." This turned out to be a most unfortunate challenge, from Harrison's perspective. News of it soon spread. When a copy reached Tecumseh and Tenskwatawa, they retired to a tent to consider their response. Emerging shortly thereafter, the Prophet proudly stated that he was pleased to cause the Sun to stand still, and he would do so 50 days hence: on June 16, 1806. Further, he said that the Sun would be darkened in a cloudless sky, and the stars would come out in daytime. So dark would it be that the birds would return to their nests, while nocturnal animals would emerge from their lairs.

It seems obvious what had happened. Perhaps Tecumseh had seen one of the almanacs of the Shakers. Maybe a British agent, eager to take any chance to provoke foment and disrupt the progress of the fledgling United States, had armed the Indian leader with this piece of astronomical intelligence. Another possibility is that a party of astronomers, looking for a prime place from which to observe the eclipse, had informed Tecumseh what was to happen. The previous year, on June 26, a partial eclipse had been visible from North America, with half of the Sun being covered; maybe Tecumseh had noticed that and realized the possibilities. Whatever the background involved, to the average Indian a darkening of the Sun was to occur at the behest of the Prophet, who would thus be proven to be the agent of the Great Spirit.

At the time of this prognostication, Tecumseh and Tenskwatawa appear to have been in the region of the Sandusky River, close to the southern bank of Lake Erie. That is significant, as we will see, because by June 16 they had moved south again, to the village the Shawnee had established on the site of the defunct Fort Greenville.

THE ECLIPSE OBSERVED

At Greenville, thousands of Indians from many tribes gathered, having heard of the prophecy. Accounts handed down to us have the Prophet pointing his finger toward the Sun at just the correct time, and as all cowered in fear he appealed to the Great Spirit to remove the obstruction and let the beneficial orb again shine down upon the land. This, of course, is said to have happened precisely as he ordered it.

But there is a puzzle here. The accounts talk of this eclipse occurring when the Sun was highest in the sky, close to midday, but in Greenville the middle of the eclipse was at about 9:45 in the morning, local time. It did occur rather closer to noon for a viewer in Boston, like Bowditch, suggesting that out on the East Coast, from where the history has propagated, there may have been some later embellishment of the story after the fact.

But that is a triviality compared with what was actually seen from Greenville. The story tells of the Sun being completely blanked out and stars being seen. We can calculate the track of totality, however, and it did not pass over Greenville at all. The southern limit of the track passed some tens of miles north of the Shawnee village. If Tenskwatawa had remained where he was, close to Lake Erie, then the eclipse would have caused just the effects he had pronounced. But at Greenville the eclipse was only partial. Admittedly only about one part in 500 of the solar disk was left uncovered, and that obscuration would have been impressive in itself, but that is not a total eclipse.

The conclusion is that the stories of the Prophet astonishing his people by causing the Sun to be blotted out cannot be entirely true. The descriptions of seeing Venus, Mars, and various stars in a black sky during the eclipse must have been transplanted from the awed accounts given by viewers who were further north at the time.

Nevertheless the eclipse did provide a mighty impetus to Tecumseh in his efforts to provoke a great rising of the Indians. In 1808 the brothers moved west into Indiana, establishing a larger settlement variously called Tippecanoe or Prophetstown. There, members of many different tribes gathered to plot the eventual overthrow of the Americans and the formation of a single Indian nation. To that end, Tecumseh spent much of his time traveling widely to garner support. One day in 1811, while Tecumseh was away, Harrison approached Tippecanoe with a thousand troops. Tenskwatawa had been warned by his brother to avoid any fighting while he was absent, but the Prophet perhaps believed too much in his own propaganda and launched an attack. The Indians were routed and Harrison burnt their village to the ground.

Having had it demonstrated to them rather painfully that the Prophet, despite his claims, could offer them no protection against the Americans' bullets, the Indians lost faith in their leadership and dispersed. The long-term effects of the Battle of Tippecanoe were therefore far more significant than might be imagined from the relatively small number of Indians killed. Tecumseh was forced to throw in his lot with the British, resulting in his death in battle two years later. Tenskwatawa was obliged to move to Canada, where he stayed for more than a decade before returning to Ohio and then Missouri when all Shawnee were ordered to move west of the Mississippi. He finally died in Kansas in 1837.

One final thing worth mentioning: Tecumseh's name means "shooting star." Apparently a very bright meteor was seen at the time he was born, and so he was perhaps destined from the start to be linked with astronomical phenomena.

9

The Rocky Mountain
Eclipse of 1878

The mapping of the dark shadow, with its limitations of one hundred and sixteen miles, lay across the country from Montana, through Colorado, northern and eastern Texas, and entered the Gulf of Mexico between Galveston and New Orleans. This was the region of total eclipse. Looking along this dark strip on the map, each astronomer selected his bit of darkness on which to locate the light of science.

Maria Mitchell, professor of astronomy at Vassar College, describing the eclipse of 1878

Several total eclipses crossed North America during the nineteenth century, and each has an interesting story to be told about it, although no other had the same order of significance as that of 1806. The authorities would not let themselves be caught out again in the same way as with the Shawnee Prophet. In this chapter we will concentrate upon the eclipse of July 29, 1878.

Eclipses had entered a period of heightened scientific study. As I wrote in Chapter 5, the heyday of eclipse watching was between the 1840s and the 1930s. Having realized that the corona was some form of extensive solar atmosphere, astronomers mounted expeditions to find out what they could during the precious few minutes of totality. Elsewhere we have learned about the discovery of helium in 1868, and other secrets of the Sun that were uncovered in following eclipses.

That is not to say that eclipses were regarded solely through the cold visage of science. As people abandoned their old irrational beliefs, there was more romance attached to eclipses. Yes, they were regarded as being perfectly understandable natural phenomena, but also wonderful things to behold for their sheer beauty. An example of this new attitude is shown in Figure 9-1. The Sun is depicted as a male deity, being embraced by the female moon goddess during an eclipse. All this is watched by an array of anthropomorphized optical instruments on the Earth below.

The total eclipse in 1878 was big news, exciting the general public throughout the land. Although it was far to the east of totality, in St. Louis the local people were thrilled by the partial obscuration they saw, and they thronged around Washington University where telescopes were trained on the Sun. Immediately after the termination of the event the *St. Louis Evening Post* put out a second edition, recounting such information as had been gathered already by telegraph from the observing parties in Wyoming and Colorado. The headlines, reading as follows, used the same symbolism as that in Figure 9-1:

THE ECLIPSE.
Old Sol Obscured by the Lunar Sphere.
The Sun God Embraces the Queen of Night.
All About the Astronomical Event of the Year.

Although eclipses had firmly entered the sphere of science, rather than superstition, it does not follow that they were viewed

FIGURE 9-1. A nineteenth-century French lithograph showing a romanticized impression of an eclipse. Human-like telescopes and binoculars watch from the Earth below as the Sun and Moon meet; the eclipse is depicted as a celestial embrace between god and goddess.

from an atheistic standpoint. Indeed many then, as now, regarded the splendor of the skies as being manifestations of their religious beliefs. As one writer put it: "Science and general education have banished all the dread these events inspired. Announced with exhaustive accuracy before their coming, fear has given way to admiration at the fixed laws, the order, the harmony of God's workings, where once ignorance anticipated accident, the coming of disasters, and tokens of the anger and wrath of the Creator."

THE SAROS AND THE NINETEENTH-CENTURY U.S. ECLIPSES

Only Washington and Canada were visited by the eclipse of July 1860, and so comparatively few people watched it. In August 1869, though, a broad eclipse track swooped down over Alaska and the western parts of Canada. Crossing the border into the United States at Montana, the eclipse path covered many of the states in the Midwest before fizzling out in the Atlantic soon after straddling North and South Carolina. Although this was relatively close to the major population centers and universities on the East Coast, astronomers tended to head further west to observe it from Iowa and Illinois, because from there the Sun was higher in the sky.

This, then, stirred up more interest for the great Rocky Mountain eclipse of 1878. But before we move on to that event, let us consider the way it is linked with those in 1806 and 1860. The saros period is 18 years plus 10 or 11 days. Triple that, and add the result onto June 16, 1806. You will get the date of July 18, 1860, when the eclipse began at sunrise in the Pacific Ocean just off the coast of Washington. Add on another single saronic period, and you get July 29, 1878, the date of the eclipse that is the main subject of this chapter. Thus these three eclipses crossing the United States were all part of the same saros cycle, a sequence that began in 1535 and will continue until the 75th eclipse in the year 2888.

ECLIPSE PREPARATIONS IN THE WEST

Expeditions were sent to the west in July 1878 from many of the established universities. From institutions in New York, Rochester, Philadelphia, and Chicago a stream of astronomers issued forth, heading for the mountains to set up their instruments (see Figure 9-2). Princeton University sent a team that was reputed to be the best equipped, with the latest telescopes and spectroscopes needed for the job. The U.S. Naval Observatory in Washington, D.C., dispatched five separate groups of observers to well separated points so as to be sure that they would not all be clouded out. As it happened, Monday July 29 dawned clear, after 12 days of unsettled weather. Very few observers had their plans disrupted in any way by clouds.

The choice of location was dictated to a large extent by the paths taken by the different railroad companies. The Union Pacific line from Chicago to San Francisco running through southern Wyoming afforded one possibility. Along that stretch of the railroad, near Rawlins, there were four separate eclipse parties, sampling alternative points across the track of totality.

Another route was further south. By taking the Santa Fe railroad from Kansas City to Pueblo, Colorado, eclipse parties could get to the edge of the plains in front of the Rockies. They could then decide how far north they wanted to go, on the Rio Grande railroad passing up through Denver to Cheyenne. Most did go to

FIGURE 9-2. A temporary observatory set up in Colorado for the 1878 eclipse.

Denver or thereabouts, the eclipse providing the greatest single influx of people ever seen by that city since its foundation. Others had determined that Colorado Springs was far enough. Samuel Pierpont Langley of the Allegheny Observatory, near Pittsburgh, was later to become an aeronautical pioneer: the NASA-Langley Research Center in Virginia is named for him. Back in 1878 he was reaching for the sky in a different way. Accompanied by several other astronomers he established a temporary observatory on Pikes Peak, the summit of which is over 14,000 feet above sea level. In the event their observations went well, but only after some consternation in the days beforehand, because several members of the party had to be taken off the mountain when they started suffering from altitude sickness.

As regards the general public, there was some argument in the preceding weeks with regard to whether businesses and factories should close up, in order to allow workers a chance to view the eclipse. An announcement was made in Denver the day before that all banks would close early, at 1:30 in the afternoon, 45 minutes prior to the first contact. Essentially all shops were closed by two o'clock, and the only thing one could easily buy was a piece of smoked glass for eclipse watching, boys walking the streets hawking such aids to anyone who had not had the foresight to prepare their own. As the *Rocky Mountain News* said, "The show was on the grandest of grand scales, free to all, without money and without price"—except perhaps in lost wages.

Several overseas visitors came to Colorado for the eclipse, in particular a group of half a dozen British astronomers, who were accompanied by Asaph Hall from the U.S. Naval Observatory. They positioned themselves at Fort Lyon, on the plains out to the east of Pueblo, where the Santa Fe Trail snaked down from Kit Carson. Hall had recently become internationally famous through his discovery, just the year before, of the two small moons of Mars we call Phobos and Deimos. In this party was Norman Lockyer, whom we met in Chapter 5. An avid eclipse chaser, Lockyer was the founder and first editor of *Nature*, a magazine that continues as the world's premier scientific journal. He was hugely impressed by the enthusiasm for the eclipse shown by the local people, sending home to London the following report: "As significant of the keen interest taken in the eclipse by all classes here, I may mention that on the Sunday before the event prayers for fine weather were offered in all the churches of Denver." Even more, Lockyer was delighted by the encouragement given to science by the U.S. government, contrasting it against the attitude he experienced back in Britain: "Strange as it may seem, this is the expressed feeling of all the authorities here, from the Chief of the State downwards. In interviews with which I have been honored, the President of the United States himself, the Secretary for War, General Sherman, and other members of the Cabinet have one and all insisted upon the importance of securing records of all possible natural phenomena, and expressed their gratification that such records have been secured in the present instance by Government aid."

SCIENTIFIC INVESTIGATIONS

For the professional astronomers, the main subject of inquiry was the solar corona. The nature of that structure, seen only during an eclipse, was as yet unclear. What is it made of? Does it contain yet more unknown chemical elements, like helium? These are the sorts of questions we addressed in Chapter 5. Back in 1878, scientists had real quandaries with understanding the corona. These are nicely encapsulated by the following passage that appeared in the *Boston Globe*: "The corona is not a solar atmosphere in the sense in which the word is usually understood, since the great comet of 1843 passed through 300,000 miles of it at the enormous velocity of 350 miles per second without suffering visible damage, or being in the least retarded; yet shooting stars passing through the upper portions of the Earth's atmosphere are completely vaporized, although their speed never exceeds fifty miles per second. What then is the corona?"

The problem with the corona was based upon its dimensions. Although its measured size alters from eclipse to eclipse due to variations in solar activity, typically it stretches up above the photosphere by a distance of the same order as the radius of the Sun. That means it is huge. The atmospheres of the Earth and Mars were known to be of very limited extent, only tiny fractions of the radius of either planet, and so astronomers were at a loss to explain the vast corona they saw. Even given that they knew that the solar photosphere was very hot, many thousands of degrees, still they could not explain the corona's extent. The calculations indicated that if it were as hot as the glowing solar surface, then the corona might extend out into space for a few hundred miles—but not half a million miles.

It was known that the corona must be very tenuous, else it would be apparent at other times than during eclipses. Limits on its thickness, in terms of the amount of light it absorbs, were set in 1878 when an observing team from Chicago noted that they could see one of the stars in the constellation Cancer right through the corona. Thus although it appears bright during an eclipse, the corona must contain very little matter, else it would have absorbed the star's light.

The eventual solution came when it was realized that the temperature of the corona is measured not in thousands of degrees, but in millions. In 1878 this was entirely unsuspected. Various specific experiments were planned to probe its nature during the eclipse. By using spectroscopes it was hoped to identify new constituents. Using polarimeters, which enable the polarization of the light received to be determined, astronomers hoped to be able to identify whether there was dust within the corona. Nowadays it is known that there is an enhancement in the density of interplanetary dust near the Sun which may be investigated by using the sunlight it scatters, and this is termed the F-corona, but a century and more ago there was scarce knowledge of such things.

Regarding the temperature of the corona, an attempt was made to measure its heat directly. These were still early days in infrared astronomy and very little was understood. But the right man to have on that task was Thomas Alva Edison, the great inventor.

EDISON'S OBSERVATIONS IN WYOMING

In 1878, Edison was 31 years old. The year before he had invented the phonograph, which is widely regarded as the second best of his ideas, behind the incandescent electric light bulb. And there is evidence that the development of the light bulb was connected with the eclipse.

Edison had been working hard on a gadget known as a tasimeter to use during the eclipse. The basic concept of this device involves a small solid block having light shone on it from a certain source in the sky, selected by using a telescope with a screen or slit arranged such that only light from that source reaches the block. Any slight temperature change produced by the incoming radiation will cause the block to expand or contract, and the stress induced is a sensitive measure of the temperature variation. That stress or pressure can be measured electrically. It was reckoned that Edison's tasimeter could show a change of just one part in 50,000 of a degree Fahrenheit, and maybe even ten times better using an improved galvanometer. The idea was to try to measure the infrared radiation emitted by the corona, and so deduce its temperature.

Edison decided to combine his expedition to the eclipse with a month-long vacation in the western states. Leaving New York on July 13, he reached Laramie a week later and stopped to buy some hunting and fishing equipment. His spot for observing the eclipse was a hundred miles west, in Rawlins, Wyoming, along with various other parties who had used the Union Pacific railroad.

Unfortunately Edison made the mistake of setting up his equipment in a chicken coop. As the obscuration of the Sun progressed the chickens decided it was time to return to their boxes, getting under his feet at the critical stage, limiting the observations. He certainly got a reading, but did not have time to do much else. Edison should have read the Chinese annals of the thirteenth century: "The Sun was eclipsed; it was total . . .The chickens and ducks all returned to roost. In the following year the Sung dynasty was extinguished."

At the start of August Edison headed onwards to San Francisco, also visiting Yosemite and various mines in the region to look into their ventilation and lighting requirements. He returned to Rawlins two weeks later to do some fishing, before heading on to Chicago and St. Louis. There he presented a paper on the tasimeter to the American Association for the Advancement of Science. On August 26 he returned to his laboratory in New Jersey, and the following day began his experiments on his electric light bulb. What is the connection with the eclipse?

In all Edison and his team of researchers tested something like six thousand different materials as possible filaments for the light bulb. (It was Edison who coined the term "filament" during these experiments.) Although they had many early successes, still the filament lifetimes were limited. It was not until 1897 that they settled on cotton thread that had been carbonized as the best they could do. In 1910 Edison's rival William Coolidge realized that a microscopically thin tungsten wire was much better, and that is basically what is still used in light bulbs today. In any case Edison's technique was limited by his insistence on using direct current, rather than the standard alternating current that soon took its place when the advantages were recognized.

Stepping back in time, Edison broke a bamboo fishing rod while angling near Rawlins and that night he threw it on the campfire. As he watched it burn he noticed how individual strands of the wood glowed white as they burned fiercely, and that convinced him that bamboo might be the best material to use for his light bulb filaments. His final solution, carbonized thread, is not much different.

LOOKING FOR VULCAN

In Chapter 13 we will be examining another type of eclipse, a phenomenon known as a *transit*. This is when a planet or some other small body crosses our line of sight to a larger celestial object. Examples are transits of Mercury and Venus across the face of the Sun (Figure 13-1), or by the Galilean moons of Jupiter across the disk of that planet (Figure 13-5).

As will be described in more detail in that section, there was a problem in the nineteenth century with astronomers' observations of Mercury. The motion of that planet appeared to be discrepant, and a suggestion for the origin of this anomaly was that there was a small, unsighted, interior planet tugging Mercury along. That hypothetical intra-mercurial planet was labeled Vulcan, even though it had yet to be found. There had been reports that it had been seen as a dark spot cutting across the face of the Sun—in transit that is—but these claims were inconsistent and ambiguous. Many people confidently expected that it *would* be spotted during this eclipse. When the Moon hid the bright solar disk it might be possible to see this faint body, orbiting close to the Sun. At least that was the idea. The *Boston Globe* began a report that morning by stating: "This is the day when the inhabitants of a goodly portion of the American Continent are to be favored with the rare pleasure of an unobstructed view of Vulcan." Several of the astronomers who trekked to Colorado and Wyoming did so specifically with a search for Vulcan in mind. Just think of the fame that would be attached to the discoverer of a new planet. Wishful thinking led to a couple of claimed sightings, but in the end it all came to naught. The reason why will be discovered in Chapter 13.

THE POPULAR VIEW

Dozens of astronomers came to see the eclipse, but there were thousands and thousands of people outside to watch it. In Denver the tops of all the taller buildings were festooned with ladies and gentlemen fighting for what they imagined to be the best spots, although the middle of a road or field would have done just as well.

This is what the *Rocky Mountain News* had to say the next day: "While the professionals, with their sails trimmed, calmly awaited Luna's approach, the average citizen was frantically engaged in hunting pieces of broken glass in the back-yard and burning it and their fingers over a dubious light on the kitchen table. The stock of street vendors of the dusky article was soon exhausted, and the demand continued up to the first moment of the contact." In this regard, some people had a little luck. The year before a phenomenal hailstorm in Iowa had smashed over a thousand greenhouse windowpanes, much to the owners' distress. Now that broken glass was proving to be a saleable item, conveniently chopped into the appropriate size for smoking and eclipse viewing.

As the partial phase progressed, the temperature began to drop. Although the day was clear, it was hot and humid. Just before first contact the temperature displayed by a thermometer left in direct sunlight was 114 degrees Fahrenheit. As the eclipse reached totality that had fallen to just 83 degrees.

In Denver totality began a few seconds after 3:29, and lasted for 2 minutes and 40 seconds. Even if your watch was not quite correct, it was easy to see when the eclipse would arrive. The track brought the shadow over Longs Peak, towering above the horizon to the northwest at a distance of about 60 miles. The edge of that shadow was moving at around 32 miles per minute, and so it took just about 2 minutes to sweep down from the heights of the Rockies and reach Capitol Hill in Denver, where thousands were massed to see it.

As the Sun went dark the corona became visible to all, although not much of a chromospheric display was seen, with only one notable prominence, perhaps two close by each other. There were stars to be seen, however, which is a staggering thing to an inexperienced observer. Regulus and Procyon, along with the Gemini twins Castor and Pollux, were obvious. Even brighter was Venus, and Mercury was seen, too.

Not everyone was so pleased with the eclipse. It was reported that the workers in the Chinese laundries went outside and "beat their gongs all through the totality." Even less happy was one lad whose sad story was told by the *Rocky Mountain News*: "The young man whose customary siesta yesterday extended beyond the period of totality, his landlady forgetting to awaken him, was around looking for a cactus to sit down on last night."

THE ZULU WAR ECLIPSE

After the great Rocky Mountain eclipse of July 1878, the next one was annular, visible as such on a line passing over Africa on January 22, 1879. It is only the partial eclipse as seen from the southeastern segment of the continent that is of interest to us here.

This was the time of the Zulu Wars, when the British were trying to wrest from the native peoples the region of South Africa now known as KwaZulu Natal. One particular battle in those wars, occurring at Rorke's Drift, is well known to many because it was the subject of the 1960s movie *Zulu* (starring Michael Caine and Stanley Baker). At Rorke's Drift, a Swedish mission station far from civilization, 100 British soldiers held off an attack by 4,000 Zulu warriors. The highest-ranking medal awarded for bravery in the British armed forces is called the Victoria Cross: at Rorke's Drift more of these were won than in any other battle in history. Seventeen of the defenders died; countless attackers also perished.

The Zulus who attacked Rorke's Drift, beginning at about three o'clock in the afternoon of that January 22, had sped there from another battle just completed nearby at a rocky pinnacle called Isandlwana. The outcome was not so favorable for the British; in fact it is often cited as being the greatest disaster in British military history. Only a few escaped unhurt from a contingent that had been surrounded by over 20,000 Zulu fighters. Of the 1,700 men on the British side, 1,329 were killed. That said, the number of Africans who died under the rain of bullets was estimated to be about 3,000.

The progress of that ferocious battle was affected by the eclipse, leading to it being remembered as the Day of the Dead Moon. In fact, because of the Zulu superstition about the state of the Moon, they had not intended to fight on that day. When the Moon disappears near conjunction they believed there were evil spirits in the air, and so they were waiting for the new moon to appear the following day. When a detachment of British troops blundered into the Zulu army, hidden from view in the undulating terrain, a spontaneous attack began, with their opponents being forced back towards the stony outcrop that is Isandlwana.

The Zulu commanders had gathered together a massive army, the warriors being deployed into a formation known as the "Buffalo Head." That is, there was a main central contingent of men, but with two horns to the sides. In this case the distance between the tips of the horns was huge, about five miles. The tactic then was to advance on the enemy, and let the horns wrap around each side, meeting at the rear to cut off any retreat by their opponents. This formation, on its grand scale, is what the British saw from Isandlwana, advancing over the horizon towards them shortly before noon.

Unlike the British, with their heavy clothes, guns, ammunition, and other equipment, the Zulu soldiers were able to move quickly on foot. Very rapidly most of the British troops who had any chance of escape dashed from them. And then, soon after one o'clock, the eclipse began, as if it were a divine sign to the Zulu that they should massacre the foreigners. At the location of the battlefield, the eclipse reached a maximum at half past two, with two-thirds of the Sun being covered.

The significance of the eclipse here is not that it hid part of the Sun, but that it made visible, as a silhouette, part of the Moon.

The Zulu initially did not want to fight because of the bad portent represented by the Moon not being seen at that time in the month. The solar eclipse, paradoxically, made the Moon obvious in the sky, giving great heart to the Africans. The eclipse was still in progress as they stormed down towards the Buffalo River to begin the assault on Rorke's Drift.

Did the Zulu know in advance about the impending eclipse? Unlike the case of the Shawnee Prophet and the eclipse of 1806, there is no evidence of any prior knowledge on the part of the Africans. The British officers, however, could and should have known about it. If they had studied military history, they would have known that it is often a good thing to avoid an engagement during or soon after an eclipse, of any variety. They might even have used it to their advantage. But that's not what happened. The Battle of Isandlwana remains one of the worst reverses the British ever suffered, although the role played by the eclipse is often neglected.

10

The Great New York City Winter Eclipse

There is no natural phenomenon that grips the imagination and stirs the soul of mankind as does a total eclipse. We ought not look at it with the eye of a dog and bark because we do not understand it. Nor ought we to look at it with the eye of a hen and tuck our heads under our wings and go to sleep because we are not interested. We must look at it with the eye of the mind.

From a pamphlet written for watchers of the 1925 eclipse

After the Rocky Mountain eclipse of 1878, viewers in the western states did not have long to wait until their next opportunities. In January of 1880 a narrow track entered California just to the south of Monterey Bay, and then passed over Nevada and northern Utah before expiring in the southwestern corner of Wyoming. On the first day of 1889 a broader track again arrived over California, this time to the north of San Francisco, then crossed northern Nevada, southern Idaho, and the northwest of Wyoming before passing over parts of Montana and North Dakota, just reaching beyond Lake Winnipeg at sunset.

Although that was it for the west for another few decades, the Deep South of the United States got an eclipse in 1900. On May

28 a total eclipse path started in the Pacific Ocean not so far off the Mexican coast, passing over that nation before clipping Brownsville, Texas, as it moved out into the Gulf. It hit land again in Louisiana, crossed southern parts of Mississippi and Alabama, and then swept fairly centrally over Georgia, and South and then North Carolina. It then departed into the Atlantic Ocean from the southeastern tip of Virginia near the town of Eclipse, which lies just across the James River and Hampton Roads from Newport News. That aptly named town gets another mention, and another eclipse, in the next chapter.

This eclipse breezed by New Orleans and many population centers along its path, and so it stirred great public interest. Many newspapers published maps showing the track of the eclipse, one of these being shown in Figure 10-1. The map is interesting for its several quirks. One is the liberty that was taken by the cartographer with various state boundaries: look at the Florida panhandle, for example. Another is the insult to the people of Illinois through the way in which their state's name is spelled. The choice of towns to mark does not seem to be consistent (between state capitals and largest cities), although I am sure that the residents of Grafton, West Virginia, were pleased to be highlighted.

INTO THE TWENTIETH CENTURY

In 1905 and 1914 there were eclipses that could be viewed from parts of Canada, but no totality in the United States. In 1918, though, there was an eclipse that swept over a dozen states in all, starting at the southern coastline of Washington and ending as it left Florida. This was on June 8. Locations further west were favored for astronomical observations, because by the time it reached

PATH OF THE TOTAL ECLIPSE.

FIGURE 10-1. A map of the track of the total solar eclipse on May 28, 1900, as published in many newspapers of the time.

the Atlantic it was near sunset, fizzling out just east of the Bahamas. On the other hand, the chance of rain is higher on the west of the Cascade Range, and so sites in eastern Oregon and Idaho were picked as being the most likely to have clear skies. In Figure 4–5 we saw the temporary observatory constructed at Baker, Oregon, specifically for this event.

On September 23, 1923, an eclipse track skimmed down the coast of southern California. Santa Barbara, Los Angeles, and San Diego were just on its edge. The obvious place to be was the Channel Islands, if you lived near L.A. and had a yacht. The track then bisected Mexico before passing into the Caribbean over the Yucatan Peninsula.

The famous New York City winter eclipse was just 16 months later. The great eclipse of January 24, 1925 began at sunrise to the west of Lake Superior. Cutting a swathe over frigid Wisconsin, Michigan, and western Ontario, it crossed the Niagara Falls and Buffalo as it entered New York. Shaving northern Pennsylvania it next darkened Connecticut and southern Massachusetts, including Nantucket Island, as discussed extensively in the next chapter. Our interest here concentrates on New York City though. As you can imagine, it caused some panic among the superstitious. But before considering what happened in the Big Apple, let us discuss some background astronomy.

PRECISE ECLIPSE TRACK CALCULATIONS

Modern science and computers have allowed the calculation of precise times and paths for eclipses, and so we know exactly when and where one should travel to experience totality. We might take pause to consider just how accurately we need to know the eclipse *time* so as to be best positioned.

The Moon's shadow sweeps across the Earth at between one-third and one-half of a mile per second (1,200 to 1,800 mph). That's the speed along the ground track, typically 60 to 100 miles wide. Taking into account the track orientation, an error of a single second in reckoning the instant of the eclipse could shift it east-west by a quarter-mile, or one part in a few hundred of its width. That's a rough estimate, but it's in the correct ballpark.

Looking back at the timings of eclipses 70 or 80 years ago, pre-eclipse predictions were often found to be out by maybe a dozen seconds, translating into a few miles as the shift in the path of totality. Analyses of historical eclipses like that of 1715, and much earlier, have been possible only since we developed the capacity to compute them with a precision rather better than a mile. Recall that Edmond Halley's predictions for 1715 were out by just 3 miles for the northern edge, but by 20 miles for the southern. Another example is the 1878 eclipse in the western parts of North America. The astronomical almanacs prepared independently in those days by astronomers at the U.S. Naval Observatory in Washington, D.C., and the Royal Greenwich Observatory in London, had maps of the track that differed by 4 miles at its borders.

Because of this timing problem it was not at all certain where the edge of the eclipse track in January 1925 would pass. One way to improve knowledge of such things was by obtaining accurate measures of when totality reached different points along the path. To this end Bell Laboratories set up a telegraphic ring of stations within the track, recording signals from them on a common chart so as to ensure uniformity. That chart is shown in Figure 10-2.

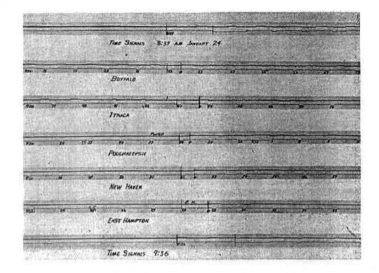

FIGURE 10-2. A chart of the time signals marking the beginning and end of totality as transmitted by telegraph, for Buffalo, Ithaca, and Poughkeepsie in New York and New Haven and East Hampton in Connecticut.

The other way of determining the edge of the track is obvious: have people spread out across the possible limits as estimated beforehand, and this is something to which we will come shortly. For the present, though, let us stick with eclipse timings. One matter that springs to mind would be the effect of the introduction of leap seconds, the need for which we discussed in Chapter 6.

Consider, for example, the next eclipse to cross the continental United States in August 2017. The track of that eclipse has been calculated already, and it is shown in Figures 15-7 and 15-8. In the decade and a half between now and then it would be anticipated that about ten leap seconds might be inserted, shifting our clocks. But will they shift the eclipse path?

The answer, of course, is no. Leap seconds are inserted only for human convenience, and eclipse phenomena are computed using astronomers' dynamical or ephemeris time systems. The leap seconds alter the time at which the eclipse will occur as displayed on a clock, but not the absolute time or the path followed.

You may think, then, that my question was misleading, but there is an important point that follows from this thought process. Although leap seconds themselves do not affect eclipse tracks, the phenomenon that makes leap seconds necessary *does* cause shifts in such tracks. Think back to Figure 6-2: the slowing of our planet's rotation rate moves eclipse paths from those that would occur if the Earth spun at a constant rate. It doesn't, because tidal drag slows it down, and leap seconds represent our solution to the problem, given the desire to keep the second a constant interval of time for various technological reasons. However, it is not possible to know ahead of time precisely how much the Earth's spin will slow before 2017.

It follows that the prediction of eclipse paths cannot be an exact science. In writing computer programs to delineate the track for some eclipse, an assumption must be made that the terrestrial rotation rate will continue to behave as it has done in recent times (and it has *not* decelerated uniformly over the past several millennia: we know that from eclipse records). We can monitor the spin of the planet on a day-to-day basis, and know it to be erratic, but the deviations from the overall trend are not huge. The derived peripheries of the eclipse track predicted a year or so ahead of time will not be off by more than a handful of yards. In consequence the argument might be considered moot because most observers will anyway be aiming to position themselves as close as possible to the central line. Recall, though, what was written in Chapter 8 concerning the desirability of being located nearer its edge.

Over extended periods the errors in the predictions enlarge. Until the time gets close, we cannot know the spin phase of the Earth at any specified juncture in the future. One may compute eclipse tracks for a century hence, but these are predicated upon an assumption that the day will continue to lengthen at the present rate, and it is virtually certain that this will not be the case. The fact of the eclipse is known, because the relevant orbits are determined with the necessary precision, but precise tracks of totality cannot be stipulated more than a century or so into the future.

The situation is analogous to flying a paper airplane. Especially given some experience one can predict with some confidence the path it will take in the inch, the foot, and maybe even the yard after it departs your fingertips. After that, who knows? Similarly there is a limit to the forward planning of eclipses, but on the scale of a human lifetime they can be predicted well enough for you to know precisely where you should be to see totality in 2045, say.

NEW YORK CITY, 1925

Let us step back now to the New York City eclipse. This was not the only major event in U.S. history to occur on January 24, 1925. In Chicago, apparently oblivious to the celestial spectacle occurring above their heads, gangsters were involved in an ongoing fight over control of gambling, illegal distilleries, and brothels. As he left his apartment block that day, the boss of the leading gang, Little Johnny "The Brain" Torrio, was ambushed by his rivals. Hit by four bullets and several shotgun blasts, he was wounded in the chest, stomach, and arm. Torrio spent ten days in the hospital fighting for his life, guarded by 30 of his mobsters. Although he recovered, Torrio decided that the millions already stashed away were sufficient for a comfortable early retirement, and he returned to Brooklyn, where his parents had brought him from Italy at the age of two. Back in Chicago, leadership of his gang was taken over by his second in command, the notorious Al Capone.

Far away in Los Angeles the first simplified traffic code in the United States was introduced. As any visitor to that city knows, a car is a virtual necessity. As the 1920s progressed, an automobile changed from being a rich man's luxury to a common means of transport for the less wealthy. As a result the rules governing the use of the road—both by drivers and by pedestrians—had steadily grown until they covered a bewildering 134 pages of turgid text, and so hardly anyone bothered to read them. On January 24, 1925, all this changed. A greatly simplified and shortened code was brought in, crammed into just four pages. This introduced, for example, the "right turn on a red light" law and also the concept of jaywalking. The public was forced to pay attention as every radio station was instructed to read the same description of the rules at eight o'clock every evening for the first week after it came into force. The ordinance was an immediate success: pedestrian deaths fell from 73 in the year preceding its introduction to 46 in the following year.

In New York, of course, cars ground to a halt as the sky darkened and the total eclipse neared. It was shortly after nine in the morning. The temperature hovered around zero Fahrenheit (minus 18 Celsius). Others would later recount just how cold they felt. Barbara Rider was one of them. In 1991 she recalled this dramatic moment in her early life six and a half decades before, when she was just eight years old: "A total eclipse took place right above Van Cortlandt Park in New York City. We got up at four o'clock in the morning to take the subway ride to this huge park in the Bronx in order to view this heavenly phenomenon. It was a marvelous display of an orderly universe and a never-to-be-forgotten experience of eerie beauty and magnificence. Also, it was bitterly cold, and my feet were almost rooted to the ground, immobile and without sensation until I tried to move. Hot chocolate in a nearby restaurant started the blood moving once more." Barbara was fortunate in that she had been taken well into the eclipse track, a good way north of Yankee Stadium. Further south, not everyone on Manhattan Island necessarily fared so well.

THE TRACK EDGE OVER MANHATTAN

In the year leading up to the 1925 eclipse astronomers knew that the border of the track would slice through New York City, but they were not sure precisely where. Looking just at Manhattan, it was clear that the track limit would pass near Central Park, but as to whether the absolute edge would be to the north or the south of it was another question.

First, let us think about the orientation of the track. It approached the city in a broad arc from the extreme west of New York state, and so it was angling down from somewhat to the north of due west. It happens that the crosstown streets in Manhattan have a similar orientation, tilted by about 30 degrees away from a precise west to east line. This meant that the track path and the cross streets were close to parallel. That was a fortuitous thing, because the street number where any observer may have been stood would on its own give a good indicator of the track limit.

Back in Chapter 3 we met Ernest Brown, an astronomer at Yale. Like most such professionals he planned to observe the eclipse from near the central line, and of course New Haven, Connecticut, was well positioned for that (see Figure 10-3). Brown, though, very much wanted to know where the edge of the track lay because this would assist him in his research on the theory of the lunar motion. Therefore he appealed to all New York residents to note where they were—on the roof of their apartment building, for example, or at a particular road intersection—and report whether they saw a total eclipse or not. In addition, observers

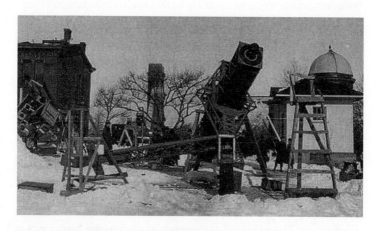

FIGURE 10-3. Snow surrounds the instruments set up on the Yale University campus to observe the January 1925 eclipse.

were stationed at every other intersection along Riverside Drive between 72nd and 135th Streets.

One might imagine this was unfair in that Brown was in effect asking people *not* to go into the zone where totality was assured: north to Yonkers or White Plains. The fact is that many of them in any case could not afford the time off work that would entail, or the traveling expense. Certainly it is true that many people south of 80th Street were disappointed not to witness totality. To the contrary, however, those very near the periphery of the track—even those slightly beyond it—actually saw a far more startling set of phenomena than those close to the center line, some tens of miles north.

Elsewhere we have described the diamond ring effect and how you are more likely to see this spectacle for a prolonged period if you happen to be located near the track edge. That is precisely what occurred in 1925. In fact the term "diamond ring" used to describe this appearance was coined by New Yorkers, when journalists asked them to describe what they had seen in their own words. It has since passed into the general vocabulary of eclipse watchers.

The result of the eyewitness accounts did enable Brown to work out precisely where the track boundary lay. It was between 95th and 97th Streets. Although it might seem that the volunteer observers had sacrificed themselves for Brown's experiment, in fact they got a far better experience than he did, even if his period of totality lasted longer.

There was great excitement, then, that Saturday night after the eclipse had passed by. Nowhere was this more so than in Chinatown. The eclipse that morning was when the Moon was at conjunction. Thirty-three hours later, as the Moon set in the west, it would be visible as a slender crescent. That meant that January 24 had been predetermined as the last day of the Chinese year, the Year of the Rat, heralding celebrations for the New Year beginning the next day.

AIRBORNE OBSERVATIONS

Quite apart from tying down the southern extent of the eclipse track by making use of the grid pattern of Manhattan and the thousands of people who reported what they saw, other important scientific results were derived from the eclipse through the cooperation of the public. A couple of years previously the American Astronomical Society had convened a "Subcommittee on Measurements and Public Cooperation of the 1925 Eclipse," suggesting a variety of ways in which the average person could contribute useful information. This resulted in the eclipse becoming a scientific experiment with perhaps the largest mass participation ever known.

The astronomers themselves were also hard at work. Many eclipse expeditions to remote locations in preceding decades had their plans upset and their hopes dashed by cloudy, rainy weather, and the prognosis for clear skies over the eclipse track were not good in 1925, given that it was the middle of the winter. As it happened, the sky was clear in New York and most other spots, which is why it was so cold. Rather than take a chance, though, arrangements had been made for instruments to be flown far above any clouds.

In all 25 aircraft carried scientists aloft, plus other interested observers who could afford the trip. The Navy airship USS *Los Angeles* flew at an altitude of 4,500 feet over Block Island, off the coast of Rhode Island and Connecticut, carrying a party of 19 astronomers plus crew. These were perilous trips. Later in the same year a similar dirigible, the *Shenandoah*, was torn apart by a storm in Ohio, resulting in the deaths of 14 passengers. The utility of airplanes was proven, however, and ten days later President Calvin Coolidge signed the act that authorized the transport of mail by air. This was in effect the first step in the process that resulted in the great airline industry of the United States.

ECLIPSES AND THEIR EFFECT ON PEOPLE

Wall Street, down near the southern tip of Manhattan, was not within the band of totality in 1925, and yet it may well have been affected. Analysts who study trends in stock market prices have suggested from time to time that surges and falls in share prices may be linked with eclipses. Apparently there is often a crash in prices within a few days of a lunar eclipse and within six weeks of a solar eclipse. It would be difficult to imagine any causal connection that could produce such a relationship and because there are several eclipses every year such coincidences may be simply a matter of chance.

Despite this there is a link with Wall Street, in one way or another. Philip L. Carret was a legendary investor and the founder of one of the first mutual funds. He watched the 1925 eclipse from Westerly in Rhode Island, from where he would also have been able to see the Navy dirigible buzzing around Block Island. Henceforth he became an eclipse fanatic, traveling around the world to see 20 total solar eclipses in all. Carret watched his final one in Barbados just a few months before his death in 1998 at the age of 101.

Not everyone is so keen. William Lyon Phelps, a professor of English at Yale from 1896 to 1933, was an astronomy enthusiast. In 1925 he was severely ill and so unable to witness the eclipse, to his perennial regret. (He made up for it by traveling to Canada to catch the eclipse on the last day of August 1932.) Phelps was staggered by the apathy he found in many others, writing as follows in his autobiography. "There are educated people who care nothing for eclipses. Some otherwise intelligent friends of mine left New York the day before that eclipse, when they could easily have waited. And another friend told me that as he and his brother (a Harvard graduate) were in exactly the right position to see it, his brother, one minute before the eclipse, said, 'Well, this is my regular time for going to the bathroom,' and went indoors. Hundreds of busy men travel six thousand miles on the mere chance of seeing what this university graduate thought quite unimportant."

Philip Carret would not have missed a chance like that. In 1979 he had an opportunity to see another total solar eclipse in the United States, this one linked to the one with which he started his odyssey in 1925. We noted in Figure 2-2 how the saros cycle pushes eclipses progressively towards the west and alters their latitudes a little. If you count three saros periods each of 18 years and 11 days after January 24, 1925, you derive a date of February 26, 1979, and the expectation of an echo of the New York City winter eclipse displaced right across the United States. And that's just what happened. On that date the track met the coast straddling the junction of Washington and Oregon, then passed over Idaho, most of Montana, and a corner of North Dakota before sweeping over the center of Canada, Hudson's Bay, Baffin Island, and Greenland.

There were a few other eclipses in North America during the

twentieth century, but not many. In February 1943 one ended at sunset in Alaska. Another in July 1945 began at sunrise in Idaho, then crossed Montana and Canada, a similar path to that in 1979 mentioned above. In June 1954 another began at sunrise in Nebraska, zipping over Minnesota, Wisconsin, and Lake Superior, and then into Canada again. We will meet a few others of interest in the next chapter.

11

Nantucket,
the Astronomically Blessed

Yf they say the mone is blewe
We must believe that it is true.

A rhyme that first appeared in 1528, from which
the commonly used phrase "once in a blue moon" is derived

ow often may total solar eclipses be seen? We might say,
with some appropriateness, once in a blue moon. That
saying is usually taken to have the meaning "hardly ever"
or "very infrequently." How that phrase came into common usage
is an example of modern folklore, with a New England connection, as we will eventually learn below.

Even with rare events like eclipses it is possible to beat the
odds. That is the subject we are going to consider in this chapter. A
quick example to begin. You will recognize from what has gone
before that a total solar eclipse might be expected to visit some
random point on Earth only every few centuries. Now think about
the miners who went to the Yukon Territory in the nineteenth-
century gold rushes. Some of them, down on their luck, may have
ended up in the northwestern corner of British Columbia, perhaps fossicking for the precious metal in the Stikine River around

the settlement of Telegraph Creek. At the end of July 1851 they could have witnessed such an eclipse. Then again, the August 1869 eclipse would have swept over them, followed by that of July 1878. That makes three totals in less than 30 years. Each occurred at the height of summer, when the chances of clear skies were greatest (indeed the dates are within a spread of just ten days). That the nearest mountains were titled the Spectrum Range seems almost too much of a coincidence. Each of those eclipses, the latter two especially, also passed over at least some part of Alaska, as if to celebrate the purchase of that state by the United States from Russia in 1867.

Back in the nineteenth century this was a pretty remote and inaccessible part of North America. Let's look instead at a particular spot a bit closer to the major population centers. For reasons that will soon become apparent, I'll pick Nantucket Island, a dot in the Atlantic merely 14 miles wide, lying just south of Cape Cod. This is a unique place: it is a town, a county, and also an island of course. Actually, it is the only locality in the United States to be so defined. On top of that, Nantucket Island as a whole is classed as both a State Historic District and a National Historic Landmark. But I choose it as a subject for consideration because of its astronomical connections.

MARIA MITCHELL, NANTUCKET'S GREAT ASTRONOMER

The quote concerning the eclipse of 1878 with which Chapter 9 began was from the writings of Maria Mitchell, who was Professor of Astronomy at Vassar College (in Poughkeepsie, New York) from 1865 until 1888. She died the following year in Lynn, Massachusetts, but she had been born in 1818 on Nantucket, and she is still strongly associated with the island.

Mitchell was a woman famed around the world, with due cause. The list of her achievements is phenomenal, especially in the context of her times, when almost all spheres of public life were entirely the provinces of men. She was the first female professor of astronomy, and indeed one of the greatest American scientists of the nineteenth century. In 1848 she was elected the first woman member of the American Academy of Arts and Sciences (and it was 1943 before another woman was voted in, a situation Mitchell herself would have deplored). Two years later, in 1850, the American Association for the Advancement of Science also admitted her to its ranks. In 1869 the American Philosophical Society accepted her as a member, again breaking ground for the female sex. She was presented with numerous awards by a variety of foreign scientific societies and governments. In fact it was a gold medal from the King of Denmark that thrust her into international prominence.

Maria was fortunate to be born into a large Quaker family, in which the parents encouraged the girls as well as the boys in their education and intellectual pursuits. As a result she became first a schoolteacher, and then a librarian, with the time to read books on astronomy and other areas of science. Her father, a cashier at the Pacific Bank on Nantucket, built a small observatory on the roof of their house, adjoining the bank building. This he equipped with a small refractor: a lens telescope with an aperture of four inches. It was installed chiefly for him to collect observations of the positions of stars on behalf of the U.S. Coast Guard. But his daughter also put it to good use (see Figure 11-1).

While scanning the skies on the night of October 1, 1847, Maria came across a comet that was not shown in any of the most recent astronomical information available to her. Her father

FIGURE 11-1. Maria Mitchell discovers her comet in 1847.

immediately wrote to William Bond, professor of astronomy at Harvard University, concerning his daughter's discovery. Bond in turn communicated the news to European astronomers, knowing that the Danish king had been for some years offering an award for any comet discovery made through a telescope. Until that time all comets had been found using just the naked eye, and the use of telescopes in searching for new ones was in its infancy.

Of course the delay in delivering a letter across the Atlantic Ocean was considerable in those days, the first transoceanic telegraph still decades in the future. As a result, before the claim reached Europe the same comet had been independently spotted by an accomplished Jesuit astronomer working in Rome, Father Francesco de Vico, and the decision had been made to award him the prize. It was quickly realized that the priority lay with Mitchell, because she had seen the comet two days earlier than de Vico, and so it was arranged that another gold medal be presented to Mitchell a year later.

This brought Maria immediate fame, and her many astronomical attainments were quickly recognized. Soon she was offered a position at the U.S. Nautical Almanac Office, carrying out the complicated celestial calculations that needed to be done by hand in this era long before the first electronic computers were built. In 1856 she began an extended visit to Europe, meeting many prominent astronomers there, and after her return she was appointed to the faculty at Vassar, where she did her utmost to encourage the female students.

The 1878 eclipse was not the only one that Mitchell witnessed. Already in 1869 she had headed to Iowa to make observations of that event. In the later year she took with her as assistants not only her sister, Mrs. Phebe Kendall, but also four Vassar graduates, all of whom had specific assigned duties to ensure that the maximum scientific benefit could be derived from their little eclipse camp near Denver.

As it happened, after a tortuous trip by rail their efforts were almost negated not by clouds, but by the railroad companies, which managed to lose their trunks in Pueblo, where they had changed lines. If those trunks had contained only clothes it would not have been a great problem, as it is easy to find new vestments, but Maria had packed the lenses from her telescopes among the soft materials to ensure they were not damaged along the way. In the end the trunks were found and delivered to Denver, and under a clear blue sky this all-female eclipse party made a series of good measurements. The only hiccup during the actual event was when one of the students was so overwhelmed by the sight of totality that she wavered from her task of counting the seconds aloud, which was necessary so that the others could time their predetermined actions.

In the light of the story of the Shawnee Prophet in Chapter 8,

Mitchell's comments regarding her sight of the lunar shadow moving off southeastwards across the plains seem somewhat surprising: "We saw the giant shadow as it left us and passed over the lands of the untutored Indian; they saw it as it approached from the distant west, as it fell upon the peaks of the mountain-tops and, in the impressive stillness, moved directly for our camping-ground. The savage, to whom it is the frowning of the Great Spirit, is awe-struck and alarmed; the scholar, to whom it is a token of the inviolability of law, is serious and reverent." As we have seen, often it has been the "untutored savage" who has benefited most from an eclipse, at least in terms of the conduct of battles and wars.

OTHER ASTRONOMICAL CONNECTIONS OF NANTUCKET

The connection between Maria Mitchell and Nantucket did not end with her moving elsewhere or indeed even with her death. In 1902 the Maria Mitchell Association was established on the island, and today the house where she was born is open to the public during the summer. Nearby is the Maria Mitchell Observatory, where a variety of research projects are ongoing, in particular some designed to encourage the involvement of young women. Her spirit lives on in that regard.

Mitchell's memory lives on in other ways, too. There are various scientific awards that bear her moniker, and there is a crater on the Moon named for her. She is also remembered in the naming of an asteroid, or minor planet, as "1455 Mitchella." That object was discovered in 1937. Similarly, on the 150th anniversary of her comet discovery, in 1997, the International Astronomical Union approved the naming of minor planet number 7041 as "Nantucket," citing the connection with Maria Mitchell.

So Nantucket has historical, current, and perennial astronomical connections. But it is eclipses that are of most interest to us here, and so we must turn to the eclipse record of the island.

QUIRKS IN ECLIPSE RECURRENCES

A random spot in the Northern Hemisphere is crossed by the track of a total solar eclipse about once every 330 years, on average. It happens that the frequency is rather less in the Southern Hemisphere (but you'll have to wait until the very end of this book to find out why).

Nevertheless all sorts of statistical quirks occur. One that is pertinent both because it is indeed in the Southern Hemisphere, and also since it is in process right now, is the case of the town of Lobito in Angola. Many eclipse watchers headed there for totality on June 21, 2001, and all being well they may return less than 18 months later when the eclipse of December 4, 2002, will also pass over that town.

Similarly there is a fair-sized area of Turkey, near the Black Sea coast, that was traversed in August 1999 and will be again on March 29, 2006. That gap of about six and a half years occurs often: as we will see in Chapter 15, southern Illinois will experience total solar eclipses in both August 2017 and April 2024. Even further into the future, the Florida panhandle will get such eclipses in August 2045 and March 2052 (make a note in your diary right away!).

The current 38-year hiatus in total solar eclipses for the continental United States is unusual in the opposite sense, being a rather greater interval than might be expected for such a large target. Even with that large gap, between 1851 and 2050 there are

20 eclipse tracks touching some part of the continental United States, an average of one per decade.

In this book we have explored the way in which eclipses follow certain distinct cycles, set by the clockwork of the heavens, producing the figurative tapestry described in Chapter 3. That does not mean, though, that they follow timetables like buses or trains. You could imagine waiting in one spot for an eclipse for a thousand years, while not a hundred miles away the lucky folk get a couple within a decade.

In the previous chapter we discussed the New York City winter eclipse of 1925. The Big Apple won't get another until 2079. That, though, represents a waiting time of only 154 years, less than half the norm. On the other hand, looking backwards in time the geographical location where New York City now stands was previously crossed by an eclipse track in 1349, and not for another six centuries thereafter. Indeed, after the 1878 eclipse in the Rockies the *New York Times* complained that "there has not been a total eclipse of the Sun within a thousand miles of this City since eclipses first became popular."

NANTUCKET AND ITS ECLIPSES

The eclipses of 1925 and 2079 bring together two very different places: New York City and Nantucket Island. Because we are familiar with major astronomical observatories being sited atop remote mountain peaks, it seems peculiar that Nantucket has so many connections with astronomy. But it does, as we have seen above.

The next connection is through eclipses. That in January 1925 was well observed from Nantucket. It happens that the next total solar eclipse visible from there is also the next one for New York City. As the Sun rises on May 1, 2079, it will be in eclipse as seen from Philadelphia or Atlantic City, but one would do better to be rather further towards the northeast. Nova Scotia and Newfoundland would be best, unless you fancy Greenland in the spring, but Long Island, Connecticut, Rhode Island, or Massachusetts will do very nicely. Indeed Nantucket will be close to the central line.

Looking backwards in time, though, Nantucket provides a stark comparison with New York, at least for the past century. Ancient times indicate nothing unusual about Nantucket: total solar eclipses in 1079 and 1478, long before European settlement, and then an annular eclipse in 1831 with less than 2 percent of the Sun uncovered. Along the way there have been many deep partial eclipses (the eclipse of May 28, 1900 shown in Figure 10-1 presented Nantucket with 95 percent solar obscuration), but the fun really started with 1925.

Less than eight years later Nantucket had a near miss. On August 31, 1932, an eclipse track came down through the middle of Hudson Bay and then Quebec Province, crossed much of Vermont, New Hampshire, and the southwestern parts of Maine before skimming the Massachusetts coast. While most expeditions went northwards, some people got a good view from Provincetown, on the tip of Cape Cod, where the Pilgrim Fathers first landed in 1620.

Nantucket was a handful of miles off the southern limit of totality in 1932. In fact, the best-selling astronomy computer program I have used for many of the calculations in this book indicates that the eclipse was total in Nantucket, which it clearly was not. That indicates some inaccuracy in the input parameters, but from the perspective of eclipse viewing the question is moot in any case. On the day in question the island happened to be covered with clouds.

Residents of Nantucket were again teased in 1959 and 1963, eclipse-wise. On October 2, 1959, there was an eclipse at sunrise in eastern Massachusetts, travelling east over the Atlantic and passing just north of Boston. About 2 percent of the Sun was uncovered as seen from Nantucket. On July 20, 1963, a partial eclipse darkened all but 6 percent of the Sun.

This was all leading up to 1970. On March 7 the track of a total solar eclipse touched down in the Pacific Ocean, crossed Mexico and its Gulf, met the United States at Tallahassee (note my earlier comments about eclipses over the Florida panhandle), and then skimmed up the Atlantic seaboard. The regions of Georgia, the Carolinas and Virginia within about 80 miles of the coast were eclipsed. At the entrance to Chesapeake Bay the track went out over the ocean, but Nantucket was in luck.

One might imagine that the island would have welcomed this as providing a tourist boom, but recall that this was only a short while after the Woodstock music festival. Proposals that a similar if smaller celebration should be staged on Nantucket to coincide with the eclipse were vetoed. Nantucket was not the only place to feel this way. The natural place to hold such a festival would have been the little town of Eclipse, Virginia, which happened to be within the track. Again the concept was rejected. The would-be festival organizers ended up taking their idea offshore, chartering a cruise liner to chase out into the Atlantic an eclipse in July 1972 that had passed over Canada. That was the first in what has become a common way of experiencing eclipses.

For a low-lying island barely more than a dozen miles wide, Nantucket did rather well, then, with regard to twentieth-century eclipses. But the law of averages must be repaid somehow. Nantucket has started the twenty-first century with a statistically freakish period in which no solar eclipses at all may be seen. In the Appendix it is shown that in every calendar year there are at least two solar eclipses of some description, and there may be up to five such events. The Sun is above the horizon for any location on Earth for just over 50 percent of the time, and the lunar shadow sweeps across almost half of a hemisphere during an eclipse (Figure 2-3). Therefore one might anticipate that each point on the Earth would witness about one solar eclipse per year, on the average, the vast majority of them being partial.

In view of that, one would imagine that it would be unlikely that any spot would pass more than four or five years without having at least a slim partial eclipse being visible. Nantucket, the island that argues with averages, is now within a sequence of 13 years with nary a solar eclipse to be peeked. After the partial eclipse on Christmas Day 2000, the island's residents must wait until November 2013 for their next chance. In the meantime they will have to console themselves with the several lunar eclipses to be enjoyed, as described in Chapter 15.

SOUTH OF THE BORDER

Total solar eclipses provide all sorts of statistical vagaries. Like New York City, England's capital, London, had to wait 575 years for such an eclipse, from 1140 until 1715 (one of the eclipses we discussed in detail in Chapter 7). Jerusalem had a gap of 795 years between 1131 and 336 B.C., but including the latter event a region near that holy city was crossed by three total eclipses within 54 years in the fourth century B.C. Similar triplets have occurred elsewhere over the past several millennia—we mentioned one in British Columbia at the start of this chapter—but as you can imagine they are quite rare.

Brownsville in Texas was mentioned in Chapter 10, in connection with the eclipse of May 1900. It happens that there is another triplet due to begin in 50 years' time, covering a region just south of that town, over the Mexican border. Three total solar eclipse tracks will intersect there on the following dates: March 30, 2052; September 23, 2071; and May 11, 2078. The Laguna Madre would seem to be the prime viewing spot for our great-grandchildren to plan to moor their yachts.

ONCE IN A BLUE MOON

We come at last to the interpretation of the phrase "once in a blue moon." Often the intended meaning on the part of the speaker as the words are uttered is "seldom, if ever." But it happens that the saying has a long and mixed up history.

Over just the past few decades the astronomically defined meaning of a "blue moon" has altered, due to a mistaken belief about previous usage. This new meaning was based on the second occurrence of a full moon within one calendar month. Because there are about 29.5 days between two lunar oppositions, in a calendar month with 30 or (much better) 31 days there is a small chance that two full moons will occur. The second of these full moons started to be referenced as being a "blue moon" only during the past few decades. That is, it's a new piece of folklore. My lengthy discussion in the Preface was based on that modern meaning. I won't repeat the details here.

It is the earlier usage of the term, which is a little more complicated, that we need to clarify. As I wrote at the head of this chapter, this has a New England connection. For many years the *Maine Farmers' Almanac* would indicate the full moons that were to be regarded as being "blue," and the rule had nothing to do with calendar months. Unfortunately, times change and during the 1940s the interpretation of the blue moon rules got a little confused. Let us look further back.

The Church has labels that it attaches to each of Sundays closest to the full moons in a year, because all the moveable feasts are phased against Easter. Similarly, farmers' activity during the year can be somewhat affected by the seasonal dates of full moons. Before the advent of artificial lights on gigantic combine harvesters, having a full moon at the time that the crops were ready to be brought in out of the fields was a huge boon. Therefore it is not surprising that each full moon was given a name. These are:

Spring season: Egg (or Easter) Moon, Milk Moon, Flower Moon
Summer season: Hay Moon, Grain Moon, Fruit Moon
Fall season: Harvest Moon, Hunter's Moon, Moon preceding Yule
Winter season: Moon following Yule, Wolf Moon, Lenten Moon

But how were those seasons defined? Some people think of the seasons beginning with the first day of a calendar month, such that spring begins on March 1. Others (like astronomers) may insist that the equinoxes and solstices mark the season starts, so that spring begins with the vernal or spring equinox around March 20. This leads to seasons that are of differing lengths, because the Earth's speed in its orbit varies during the year.

The farmers' seasons, however, are defined in another way again. In essence the year is split into four equal seasons, each lasting 91 days plus a bit. That's a fairly straightforward way to do things. Three lunar months (synodic months) last for a total of 88.6 days, indicating that although three would be the norm, a fourth full moon could fall within one of those farmers' seasons. To look at it another way, each calendar year contains at least 12 full moons, but in about one year in three there is a 13th full moon. That means that there is one more to be inserted, beyond the dozen named full moons listed above. And that one is called—you've guessed it—the *blue moon*. In a season with four full moons it is the *third* of them that is termed the blue moon, according to these rules.

As a result, the blue moon can only occur in February, May, August, and November: that is, close to one lunar month before the next equinox or solstice, although those points are defined slightly differently in this scheme of equal length farmers' seasons. Consequently the blue moon by this definition can only occur around the 21st or 22nd day of one of those calendar months, and never on the 30th/31st as happens according to the recently evolved version of the meaning of "blue moon."

By this original rule, the next blue moons will occur in November 2002, August 2005, May 2008, and November 2010. There are gaps between two or three years, then, which gives you another handle on what "once in a blue moon" may be taken to imply.

I wrote that this is the "original rule," but in fact it is not so ancient in itself. Tracing through such volumes as the *Maine Farmers' Almanac* indicates that it sprang into usage in the agricultural community of New England around the middle of the nineteenth century. The couplet with which this chapter began is around three times as old as that, dating back to before Shakespeare's era. If we are to ask, "why blue?" then we need to go back a long way.

The answer to that query is that no one really knows. People may have started saying that "the mone is blewe" in 1528, but the following year a similar phrase appeared, with a different color involved: "They woulde make men beleve that ye Moone is made of grene cheese." Of course everyone knew that the Moon is not made of green cheese. The literal meaning of this piece of doggerel is similar to saying that someone would argue that black is white.

The origin of the blue moon pairing of words is the same, a straightforward example of something that is known not to be true. Yes, under certain atmospheric conditions the Moon in the sky may attain a somewhat bluish tinge, but that is irrelevant. "Once in a blue moon" is a common phrase with ancient roots, and its interpretation in terms of astronomical phenomena has changed over the last century or so. In effect, though, it still means once in two or three years, about the same frequency with which a total solar eclipse occurs in some accessible part of the globe, in fact.

WYOMING REVISITED

Nantucket is a picturesque location from which to witness an eclipse, but there are none due there soon. The next total solar eclipse to pass over the continental United States is in August 2017. In Chapter 15 the path it will take is discussed, and I suggest that the Grand Tetons might be the pick of the places from which to watch it.

Looking back in time, it happens that the eclipse tracks in

1878 and 1889 framed Yellowstone National Park rather nicely, the Grand Tetons also lying within their crossing zone. In 2017, however, the track edge only shaves the southern border of Yellowstone, leaving the Grand Tetons as the only choice up in that corner of Wyoming. If you do go there to see that eclipse, recall its nineteenth-century siblings, watched by all and sundry when the West was far wilder than it is now.

12

Eclipses of the Third Kind

Damn the Solar System. Bad light; planets too distant; pestered
with comets; feeble contrivance; could make a better myself.

Lord Francis Jeffrey (1773-1850)

So far we've looked at two basic types of eclipse: solar and
lunar. Our Sun is not the only star whose face the Moon
can pass across though. Every month the Moon, in its passage around the Earth, blocks out the light from some millions of
stars in the Milky Way, and many extragalactic objects, too, each
reappearing about an hour later behind the trailing limb of the
Moon.

Most of these remote light sources are extremely faint, but
every so often the Moon will obscure some particularly bright
star, and numerous amateur astronomers will be keen to witness
the event. The target might be Regulus, the bright white star in
the constellation Leo, or Aldebaran, the vivid red object in Taurus,
or some other familiar heavenly jewel. Nor do the planets escape
alignment with the Moon: because they occupy a restricted band
about the ecliptic, they, too, are frequently blotted out for a brief
time.

One may think of these as "eclipses of the third kind," but
there is a specific astronomical term attached to them: *occultations*.

This is an area of astronomy in which amateurs are able to make vital contributions to our knowledge base.

LUNAR OCCULTATIONS

Imagine that the Moon is due to cross a particular well-known star. What useful information can be obtained about our natural satellite?

First, because we can measure and catalog the coordinates of the stars with great precision, by timing the instant at which the star disappears behind the Moon one may determine the lunar position at that instant with similar accuracy. It is relatively easy to ascertain the locations of objects that effectively stand still, like the stars. Because they are moving in concert around the sky, a telescope can continuously track them if it is rotated at just the right rate to compensate for the turn of the Earth. Yet this is not so with the Moon or other members of the Solar System, which are in constant but variable motion relative to the static background of stars. Timing an occultation to a fraction of a second allows the observed location to be referenced against the predicted position from the computed ephemeris, perhaps leading to an update.

Nineteenth-century astronomers argued over what they saw through their telescopes when the Moon occulted a star. To many observers it seemed that the image of the star was projected onto the dark lunar disk, seeming to remain visible even after it was obvious that the star must be hidden. Some claimed that this image seemed to be colored even though the star may have been white.

Debates over this phenomenon raged for years, various hypotheses being advanced for its origin. In those days the nature of light was still a mystery. Some argued that the Moon was partially translucent, acting like a cloud whose periphery lets some light through (the "every cloud has a silver lining" effect).

Eventually it was realized that the apparition is simply an artifact of the human eye. It is similar to staring at a light globe for a few seconds, and then looking towards a dark background, resulting in a residual colored image: your retina takes a short while to recover from the bright light it had been sensing. Shakespeare knew all about this, having Katherina in *The Taming of the Shrew* say this:

> Pardon, old father, my mistaking eyes,
> That have been so bedazzled with the Sun
> That everything I look on seemeth green ...

Precisely the same thing happens if you follow a star with a telescope as it slips behind the Moon: paradoxically the stellar image seems to creep over the lunar landscape for a second or two, even though the source has already disappeared from view.

The second sort of quantitative information about the Moon that may be obtained from modern-day occultation observations pertains to its surface contours. Suppose that a particular occultation was timed by a string of observers spread over some hundreds of miles. If the Moon were exactly spherical, then there would be a simple arithmetical relationship between the times they recorded. But we know that the Moon is not spherical: rather, it is mountainous in some regions, deep canyons and rills permeating the surface elsewhere, and it is pockmarked with craters, too. Imaginary straight lines from the star to each of the observers, just touching the lunar limb, will variously strike crater rim, mountaintop, or slip through a deep valley. Because of this, some watchers will record the occultation as occurring a split second early, others a similar time late, compared to a perfectly even curve.

With a concerted effort, and accurate knowledge of the observers' positions and timings, a contour of the lunar limb may be drawn up. In addition, because the Moon vacillates slightly, not presenting a completely constant face to us, each occultation presents the opportunity to study a different arc drawn across the Moon's surface. An especially valuable opportunity occurs when a star passes virtually parallel to the lunar limb—a grazing occultation—because then observers at critical locations on the Earth see it being successively hidden and then briefly revealed as it skims along the serrated edge of the Moon. Observers separated by just a mile will see different aspects of the Moon's crinkled fringe.

STARS, GALAXIES, AND QUASARS

The above studies help us to understand the Moon itself. In the same way as solar eclipses allow the Sun's corona to be studied, so lunar occultations enable astronomers to investigate the distant light sources being occulted. That is, we can discover things about the stars and galaxies involved from the way in which the Moon cuts off their light. To understand what is going on, we must first discuss some background information about the behavior of light.

The wave nature of electromagnetic radiation (which includes visible light and radio waves) imposes a fundamental limit on the resolution or detail achievable with a specific optical system, even if that system is perfect (here perfect means that it is precisely aligned with aberration-free components, an unachievable idealistic limit). This means that there is no point in using an eyepiece on a telescope with ever-increasing magnification, even if you are in orbit on the Space Shuttle, because the resolution any telescope can deliver is limited by the laws of physics. The relevant law in this case governs the diffraction of light as it passes the edges of an opening such as a telescope aperture. (As light passes through an aperture some part of it is deviated in its path, and this is called "diffraction.")

For any optical system a measure of the best-possible resolution or resolving power (R) is simply the ratio of the wavelength (the Greek letter λ is normally used) to the diameter of the aperture (D). A factor of 70 converts the result into degrees so long as both λ and D are expressed in the same units—usually meters—so that the resolution may be expressed as $R = 70\,\lambda/D$.

An example will assist here. Consider an optical telescope with an aperture of 5 meters, such as the 200-inch reflector at Palomar Mountain in California. Observing at a wavelength of 500 nanometers (that is blue-green light) the limiting resolution, at least in theory, is about seven millionths of a degree. Imagine that the great telescope is directed horizontally at two bright, shiny pins stuck in a pincushion ten miles away. If they are separated by more than two millimeters (one-twelfth of an inch) then the telescope can resolve them as being separate, at least in principle. If they are closer than that limit, then they appear as a single object: the telescope is not capable of splitting them, even under the ideal limits cited above.

In reality any optical system is not perfect, and most importantly ground-based telescopes are used to watch astronomical objects through the atmosphere, which is turbulent and so blurs the images formed. This image degradation is what astronomers call the *seeing*; it is what causes stars to twinkle. The very best observatory sites may have seeing as good as one part in 10,000 of a degree, but that is about ten times worse than the theoretical resolution of a perfect large telescope. This problem with the atmosphere is one of the main reasons for putting systems like the Hubble Space Telescope into orbit far above us.

Now consider the implications of the finite resolving power of telescopes for our observations of even the closest stars. If these were about the same physical size as the Sun, then their disks would only appear a few millionths of a degree across because of their huge distances. Even a perfect five-meter telescope in orbit above the atmosphere, escaping its detrimental effects, would not be capable of resolving the nearest stars. For this reason the Sun is the only star for which we have direct pictures of its shape and features (although there are complicated techniques that allow profiles of nearby stars to be mapped).

One might ask then how we could measure the sizes of the stars. One answer lies with occultations. If the light from the star is fed into a detector that gives a readout of the intensity as it changes every microsecond then, as the lunar limb quickly slices across the stellar disk, the way in which the starlight diminishes will allow a deduction of the star's size. The disappearance would take about a hundredth of a second from first contact until the star is completely obscured, so that if the instrument's time resolution is good enough then one can obtain a measure of the light profile across the stellar disk. In this respect the Moon acts like a knife-edge sweeping across the sky at known speed.

Now, instead of a star being the target of interest, consider a galaxy. That galaxy might be a major fraction of a degree wide, although most are more distant and have apparent sizes only around one-hundredth of a degree. That, though, is much larger than the apparent size of an individual star. In consequence, during an occultation the total light flux collected from a galaxy drops off over many seconds of time. This means that the light signal observed during any occultation enables astronomers to differentiate between objects that are essentially points or at most very small disks, like stars, and sources that are extended, like galaxies. On the other hand a binary star (a pair of stars orbiting around each other) would produce two distinct downward slopes in received brightness: first one would be hidden by the Moon, and the other a brief instant later. The relative timing of the two decreases in the light signal would render a measure of the separation of the two stars.

In the early 1960s a new class of celestial objects called *quasars* was identified. (We will come to the origin of that word shortly.) These were unusual in that they looked small and bright, like stars in our galaxy, and yet they had huge redshifts, indicating distances from us of billions of light-years, putting them at the periphery of the universe. The *redshift* of a cosmological object is the displacement of its spectral lines owing to the Doppler effect. A familiar analog for sound rather than light is how the pitch of an ambulance or police car siren alters as it whizzes past you. There is a change from a deviation towards a higher frequency to one at a lower frequency. Celestial objects receding from us at an appreciable fraction of the speed of light have the wavelengths of their emitted light effectively increased towards the red end of the spectrum, leading to the term *redshift*. It is believed that their speeds relate to their distance, a large redshift implying a vast separation from us. It is by using this assumed distance-speed relationship that astrophysicists are mapping the universe in three dimensions.

When quasars were first recognized the initial question was

whether they are peculiar stars and nearby, or peculiar galaxies and distant. Although telescopes could not resolve their forms, occultation observations indicated that they were small but extremely powerful emitters of light, hence the name quasar: it is short for *quasi-stellar object*. Their true nature is still a mystery, in that they seem to emit far more energy from a restricted volume than is easily explicable using our present knowledge of physical processes.

RADIO OCCULTATIONS

The Moon has also been utilized by radio astronomers to investigate the angular sizes of celestial objects. We saw above that a large optical telescope has a resolving power, in principle, approaching one part in 100,000 of a degree. Radio telescopes have much bigger apertures, and there are several with diameters over 100 meters. Let us use that in our equation $R = 70 \, \lambda/D$. One might imagine that this would render an improved resolution, but the radio wavelengths employed are much longer than those of visible light, typically $\lambda = 1$ centimeter. Putting those two figures into the equation renders a resolving power $R = 0.007$ degrees, a thousand times less than that of the optical telescope. That is, the detail that might be mapped with even a large radio telescope dish is quite limited.

To overcome this handicap, in the early days of radio astronomy, when little was comprehended about the radio universe, lunar occultations were regularly employed to delineate the dimensions of newly found radio sources. The way in which the received radio signal varied in time could indicate whether emission was occurring from throughout a galaxy that the Moon happened to scan over, or only from a discrete source at the galactic center, for example.

In recent decades astronomers have used more sophisticated techniques to counterbalance the physical limitation of the resolving power achievable by single radio dishes. By linking together many small radio telescopes it is possible to obtain resolution equivalent to a single much larger dish, because the value of D to be used in the equation is given by the separation of the smaller dishes. That baseline length may be some miles, as in the case of the Very Large Array near Socorro, New Mexico. In fact, by linking together radio telescopes spaced across the whole globe, baselines of thousands of miles are feasible. The next step is to have radio telescopes in orbit, making even longer baselines and so radio maps of distant galaxies with unprecedented detail. The first step in this progression, though, was taken when lunar occultations were employed to set limits on the sizes of cosmic radio sources.

MEASURING ASTEROIDS

Although cartoonists often depict asteroids as being spherical, in fact they are mostly of irregular shape, so it is incorrect to think of them having a "radius." The major planets are spherical because of their huge masses: energetically a sphere is the form that any large body would assume, if self-gravity were the only significant factor. Without the Earth's geologically active interior, producing continental drift and volcanoes, the Earth would have no mountains and would be a solid sphere covered by continuous ocean.

For a large object to obtain a basically spherical form, the tensile strength of its component material must be overcome.

Therefore the shape attained depends on the comparative values of that strength—the ability to withstand distortion—and the gravitational force trying to pull it into a sphere. A fluid has essentially zero strength, so it attains a spherical form no matter what the size. For a solid body it is different.

In the case of an arbitrary asteroid ("minor planet" is an equivalent term), the rocks and metals of which it is composed would be strong enough to maintain an irregular shape, unless it were more than a hundred miles or so in size. There are only a few dozen asteroids of such dimensions. There are also about a million closer to one mile in size, most of them in the main belt between Mars and Jupiter. The total mass of all the asteroids in the main belt is less than that of the Moon.

The largest asteroid is called 1 Ceres, and it was the first discovered minor planet (which is why it has that preceding number one in the master list), on the opening day of the nineteenth century; it has a diameter of 580 miles. Along with a handful of other minor planets, Ceres is large enough to be resolved to some extent using the Hubble Space Telescope. These really large rocks are found to be spherical, due to their self-gravity, whereas the more-numerous smaller asteroids have all sorts of convoluted shapes (see Figure 12-1).

Small asteroids are not spherical, then, and one would like to measure both their shapes and sizes. Given that most asteroids appear merely as pinpricks of light in our telescopes, how can we fathom their dimensions? It happens that occultations enable astronomers to obtain such measurements.

If an asteroid were to pass across the face of the Sun, then we might see it in transit (as is discussed at the end of Chapter 13) but it would be so tiny that all that could be seen would be a little dark

FIGURE 12-1. Minor planet 433 Eros photographed by NASA's *NEAR-Shoemaker* satellite during 2000. Eros is about 20 miles long, but less than ten miles wide: it is obviously irregular in shape and has been struck by many smaller objects.

spot. No shadow would be cast on the Earth's surface because the Sun appears much larger than the asteroid. No measurement of the asteroid size would be possible unless it were very close to us. It happens that Mars has two moons, named Phobos and Deimos, which are captured asteroids orbiting very close to that planet. As a result they do cast distinct shadows on the Martian surface (see Figure 12-2). For an asteroid observed from the surface of the Earth, to get an effective "shadow" whose size might be measured we would need a smaller light source than the Sun, such as a star

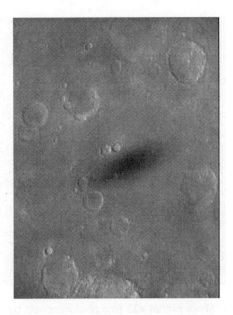

FIGURE 12-2. Phobos, one of the two moons of Mars, is only about eight miles across but it is still big enough to cast a shadow on the surface of the planet below, as in this image obtained with the *Mars Global Surveyor* satellite.

far away within our galaxy. This could produce an occultation, if the alignment were right.

Imagine that a 100-mile wide asteroid cuts across our line of sight to some distant star. We will probably not have its trajectory determined with enough precision to be sure where its shadow will pass, and as of yet we do not know its size or shape. If the movement of the shadow is west–east one might organize a team of a dozen or so observers stretched along a line north–south for

300 or 400 miles, to be sure of intercepting that shadow. Each astronomer would be armed with a small telescope and stopwatch, plus some absolute time reference such as a GPS receiver or their wristwatch accurately calibrated against standard time, and they would watch as the asteroid closed in on the star. Some would see the star blink off for a short while, as the asteroid eclipses or occults it, whereas those at the northern and southern extremes of the line would not see the star disappear at all, but just slip close past the asteroid. (Such an event is termed an *appulse*.)

The limits along the line of humans from where the star was occulted will render the asteroid dimension along that axis, perpendicular to its apparent motion. But its size in the other direction (that is, along the shadow path) and even shape may also be deduced from the observations. The duration of the occultation timed by each observer indicates the length of the star's path behind the asteroid as seen from the particular viewing location. The idea is sketched in Figure 12-3.

Because it is difficult to predict the eclipse path for an asteroid far ahead of time, due to uncertainty in its orbit, occultation chasing may be a haphazard and frantic affair. One afternoon in October 1981, while I was a graduate student at the University of Colorado, with a colleague I got a call from astronomers at the Lowell Observatory in Flagstaff, Arizona. They said that an asteroid occultation had just been predicted for that evening and we asked if we could please observe it from the on-campus observatory. This we did without any great trouble and sent off our timings. (That colleague, by the way, was Chris McKay, now one of America's most prominent planetary scientists; he works at NASA-Ames Research Center in California.) The Lowell observers had some problems though. They had found that the track was going

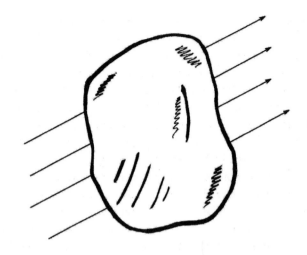

FIGURE 12-3. How the size of an asteroid can be determined from occultation data. Observers spread out over the shadow ground track and measure how long the starlight blinks off, each viewing a different path for the star behind the asteroid as shown here by the arrows. Knowing the speed of the asteroid, its dimensions along the direction of motion may then be ascertained. Any astronomer in the team who was located too far north or south would not see any occultation, and so the size of the asteroid crosswise can also be determined in this way.

to pass north of them, over Utah, and so they scrambled in their cars carrying two portable telescopes. Ideally one would organize for the observation points to be well separated so as to give the best distribution of chords across the asteroid. On campus in Boulder, Colorado, our telescope was fixed; but the mobile teams could in principle drive to locations giving an equable spacing over the occultation track. In the rush the teams lost contact with each other, and by chance the two sets of mobile observers managed to choose sites giving precisely the same chord. With the whole of the Utah wilderness to choose from, they had picked separated but equivalent points. As the final publication reported, "As a result, they were deployed in accordance with Murphy's Law."

The specific minor planet observed in that case was 88 Thisbe. The result of the analysis was that it measures about 144 miles across, around 10 percent more than the value estimated from earlier data (it is possible to estimate asteroid sizes by seeing how bright they are, and couple that with a guess at the fraction of sunlight they reflect). Ten percent in size means 20 percent in area, or 30 percent in volume and density. It was also obvious from the results that Thisbe is not precisely spherical. Clearly occultation measurements are scientifically useful.

THE MASSES OF COMETS

Asteroids are mere lumps of rock and metal, scattering tiny fractions of the sunlight impinging upon them. This, coupled with their great distances from the Earth, make them difficult to spot unless you know just where to look, using a substantial telescope. The largest, Ceres, was found only two centuries ago, even though it is almost 600 miles across. On the other hand, from the images returned by the spacecraft that were launched to greet it in 1986 we know that Halley's Comet has a solid nucleus only five to ten miles in size. It is irregular in profile, shaped somewhat like a potato. That nucleus reflects merely 3 or 4 percent of the incident sunlight. Nevertheless, for over two millennia our various civilizations have been recording the returns of Comet Halley. How could this be?

The fundamental difference between comets and asteroids, regarding their appearance in a telescope, is that comets are largely composed of ice and other volatile material that starts to sublimate as the Sun is approached. At 3 AU from the Sun, midway between Mars and Jupiter, the cometary surface heats sufficiently for water to start to vaporize, forming a tenuous cloud around the nucleus. Such a cloud—called the *coma*—may be over a hundred thousand miles across, bigger even than Jupiter, the king of the planets. Some of the gaseous products may be dissociated and ionized by the solar ultraviolet flux (water may split into hydrogen and oxygen ions, for example), and then swept outwards by the solar wind, giving comets their characteristic ion tails that seem to glow bluish. Ejected dust and meteoroids trail the cometary orbit, producing a secondary tail, usually pinkish in color. These tails may be tens of millions of miles long.

These huge expanses of fine material scatter a great deal of sunlight, which makes comets easy to see compared to dark asteroids. But how did astronomers first discover the true size of cometary nuclei, given that the only comet we have seen up close—the only one for which resolved images of the nucleus are available—is that bearing the name of Edmond Halley?

In Halley's day comets were believed to be much more massive than they really are. We now know that a cometary coma is a very tenuous gaseous shroud surrounding a tiny solid lump, keeping it from view, but in the eighteenth-century comets were thought to be huge, bulky affairs. One early hypothesis for how the planets were formed was that a gigantic comet had collided with the Sun, causing material to be ejected like the rebounding drop of liquid when a sugar cube is plopped into a cup of coffee. Individual drops were imagined to have coalesced into the separate planets. There are various problems associated with the physics involved in that idea, but in any case we now know that comets are much smaller. When they do hit the Sun, they are simply swallowed up (see Figure 5-2).

The way in which astronomers developed this understanding was through studies of occultations. Although a cometary coma looks bright, that cloud is really very thin indeed, with a density lower even than the filigree mist hugging the landscape on a warm summer's morning. Because of its vast dimensions the coma scatters much sunlight, but still it does not absorb much of the starlight coming from behind. Similarly, in thick fog your car headlights may only allow you to peer only 10 yards ahead, the water droplets reflecting so much light back into your eyes that you can see little else, but another car's headlamps can be perceived over a hundred yards away, permeating the gloom. In the same way, astronomers could probe the contents of a comet's coma by following the light of a star passing behind it. They were surprised to find that the starlight was almost always uninterrupted, penetrating the gas cloud with very little diminution. The deduction was clear: the observed parts of comets are mostly gas, originating from a tiny solid mass at the center. Comets are easily seen once the ice starts to sublimate and form that misty cloud, but when far from the Sun a comet has no coma and the bare nucleus is difficult to detect.

From more recent radar and other observations, most cometary cores are estimated to be only a mile or so in dimension. However, this smallness of cometary nuclei was first recognized from occultation investigations that, as described above, showed no occultation at all. Stars shine unabated through the tenuous but extensive comae, missing the solid nuclei.

A FUZZY OCCULTATION BY MARS

William Herschel was mentioned earlier; he was the German-British astronomer who discovered Uranus in 1781. His sister Caroline found many comets using her brother's telescopes, both from the city of Bath, where they had been living, and also from Slough, where they later moved. (Slough is near where London's main airport, Heathrow, was much later built.) The Herschel family had moved closer to the capital under the patronage of King George III. Nowadays the idea that major astronomical discoveries might be made from your rooftop or backyard in such locations seems bizarre, observatories being built on mountaintops in remote locations far from city lights, but two hundred years ago the skies were still relatively clear. The smoke of the Industrial Revolution was yet to have a crippling effect on sky translucency, and electrification causing light pollution (one of the main banes of modern-day astronomy) was an unimagined development.

If you ever visit London for the shopping, after the famous Oxford Street one of the best-known areas is Kensington High Street. Shoppers bustling along there might be surprised to learn that one of the world's largest telescopes was once situated nearby. Looking up a street directory, one may find Observatory Gardens (a road, despite the name), a few hundred yards off High Street. On that site, since built over, Sir James South established an observatory that stood for 40 years until his death in 1867. The blue plaque marking the spot is incorrect in stating that South's dome housed the largest telescope in the world. Actually it was the biggest refractor (lens telescope); Sir William Herschel, who had died in 1822, had previously constructed larger reflecting telescopes (using curved mirrors) out at Slough. South's telescope had a lens just below 12 inches in diameter. It is still in use today, at the Dunsink Observatory just outside of Dublin.

Although he has since been mostly forgotten, South was a very prominent astronomer in his day. He was one of the founders of the Astronomical Society of London in 1820 and, as the sitting President, pivotal in securing its royal patronage through contacts assembled by having the gentry come to Kensington to view comets and nebulae through his several telescopes. Thus the Charter of the Royal Astronomical Society, granted in 1831, begins with South's name. On the other hand the first Fellow of the Royal Astronomical Society could be claimed to be Charles Babbage, whom we met earlier, because he was listed first amongst the founders, due to his alphabetical advantage, being followed by Francis Baily (of Baily's beads fame).

In those days the scientific circle was limited. John Herschel, the son of William, often observed the heavens with South, and they jointly drew up catalogues of binary stars. The advent of electrification was mentioned above; this was in part due to the pioneering investigations of Michael Faraday, who frequented South's private observatory, as did Isambard Kingdom Brunel, the great engineer of the early Victorian age. Babbage was also a good friend, and it was ill-feeling fostered by a court case over the mounting of South's large telescope (which he claimed to be inadequate), that led to the opposing party recommending that the government cease all funding of Babbage's computing machines. Babbage made the political mistake of appearing as a witness on South's side in a trial that divided the scientific establishment. South was a fiery controversialist, never far from an argument with someone, and Babbage had a similarly bellicose temperament.

With his great telescope South made comparatively few

useful observations, forever complaining that its pivot wobbled, blurring the objects he wished to monitor. In 1830, though, he did make a revolutionary discovery. While watching the planet Mars moving through the constellation Leo, he saw it pass in front of a bright star.

For all his faults and intellectual limitations, South was an experienced visual observer, and he recognized that this Martian occultation was not like the numerous lunar occultations he had seen previously. Instead of suddenly blinking off (perhaps with the "projected image" effect mentioned earlier: South was one of those who had noticed this visual phenomenon and debated its origin), as Mars crept up on the star he noticed that the starlight reaching his eye slowly wavered and attenuated.

How could this be? South made the correct deduction: Mars has a substantial atmosphere. Rather than the knife-edge provided by an airless body like the Moon, Mars has a fuzzy border. This produces effects like those we depicted in Figure 2-8, when we were considering how the Moon still receives sunlight, mostly from the red end of the spectrum, during a total lunar eclipse. When watching Mars as it cut across the star in question, South saw that the starlight was gradually absorbed by the ever-thickening layer of Martian atmosphere needing to be negotiated for the light to reach his eye, glued to the ocular of his precious instrument.

Using the primitive equipment of the era, little was yet known about Mars. It presents merely a ruddy disk through a telescope, with a hint of pale colorless patches at top and bottom, the polar ice caps. The imagined canals of American millionaire Percival Lowell were still many decades in the future, along with ideas of Martians and H.G. Wells's *War of the Worlds*. From his private observatory near the heart of London, largely surrounded in those days by green fields, James South discovered that Mars has an atmosphere via his acute observations of that planet eclipsing a star. That's something to remember next time your underground train rumbles through Kensington and Notting Hill Gate, not half a mile from South's old observatory.

THE RINGS OF URANUS

When William Herschel spotted Uranus he thought it was a comet, and its true nature was not recognized for some time. When later observations indicated it to follow a near-circular orbit, not an elongated ellipse like the path of a comet, and the disk visible through suitable telescopes looked like those of Jupiter and Saturn, not the nebulous, variable form of a comet, the scientific world was astounded. No new planet had ever been identified, the naked eye planets out to Saturn having been known since time immemorial. Apart from the visits of sporadic comets, it had been assumed that the Solar System as known was complete.

To the greater glory of Britain, its astronomers tried to name the new planet the *Georgium Sidus*—George's Star—in honor of the king. (Note though that the king, like Herschel, was German in origin: the House of Hanover ruled Britain until the death of Queen Victoria in 1901, she having been prohibited from becoming the monarch of the province of Hanover by virtue of her sex.) In France and elsewhere astronomers would have none of this, and the title Uranus was eventually accepted internationally. The attempted foisting of the name George upon the planet led to regal approval for Herschel, however, and he became Royal Astronomer (not Astronomer Royal: there was already one of those), with a liberal monetary allowance. Astronomers know that it is not only stars that glisten.

In subsequent years numerous studies of Uranus were conducted, for example leading to the discovery of its several large satellites. It also became apparent that it orbits the Sun with its rotation axis tipped right over, leading to each pole having 42 years of summer followed by 42 years of winter.

Because Uranus never comes closer than about 1,700 million miles from the Earth it is difficult to investigate the planet in detail. Our best data come from the flyby of the planet made by NASA's *Voyager 2* probe in 1986. Just a handful of years before that, an occultation experiment led to a discovery that allowed the planning of some important data collection with *Voyager 2*.

Back in 1830, James South used his eye at the telescope to see Mars gradually extinguish the light from the star in Leo that he was watching. Nowadays we can conduct much more sophisticated experiments, using electronic light detectors. For example, not only will the brightness of a star be attenuated by the atmosphere of a planet, but also its position will shift due to refraction (or bending) of light in that atmosphere. This is why it takes so long for the Sun to set: refraction in the terrestrial atmosphere shifts the apparent position of the Sun as it approaches the horizon by fully half a degree. Observations of such effects in other planets' atmospheres during occultations allow astronomers to probe the density and profile of those atmospheres with a resolution many times better than otherwise feasible.

The problem is that Uranus has such a small disk that it rarely crosses stars sufficiently bright for useful data collection. Even when such an event occurs the planetary shadow is unlikely to pass over a major observatory, in the same way as a total solar eclipse is not often seen from, say, the many observatories in California, Arizona, or Hawaii. In 1977 a good occultation by Uranus was due, but to observe it a chase along the shadow path was necessary. Actually NASA maintains aircraft for high-altitude astronomical observations, and one was used to collect data in this case, the intention being to improve our understanding of the atmosphere of Uranus before *Voyager 2* got there.

But the observers got a surprise. Having switched on their equipment and acquired the star well before the occultation was due, they found that the light signal dipped not just once but several times while the star was still well separated from the planet. In itself one might explain away this as being due to some instrumental glitch, or extreme altitude wisps of terrestrial cloud, but after the planetary occultation had concluded continued data collection provided another set of dips in the signal. These were of the same form as the first set and symmetric about Uranus itself.

The explanation for these observations was clear: Uranus possesses a set of rings that had not previously been suspected. When *Voyager 2* reached Uranus it was instructed to look for the rings in close-up, with a successful outcome. The Hubble Space Telescope has since been used to get pictures of those rings, such as in Figure 12-4.

. . . AND THOSE OF NEPTUNE

Similar occultation observations involving Neptune were made during the 1980s. These also provided a hint that the planet has rings, but with a difference.

In the decades after Uranus was spotted, astronomers followed its progress in order to chart its orbit. Because that planet takes 84

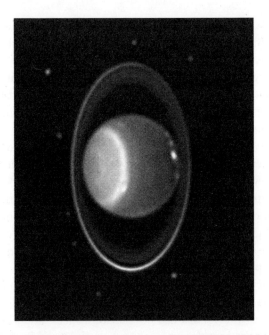

FIGURE 12-4. The rings of Uranus photographed using the Hubble Space Telescope in 1998. The rings were discovered through occultation observations 20 years before. This is the true orientation, because the spin axis of this planet is tipped over, and the rings orbit above the equator. Several of the Uranian moons can be seen, along with bright areas on the cloud-canopied planet itself.

years to circuit the Sun, less than three Uranus years have yet to elapse since it was discovered (and it is sobering to note that Pluto has not completed even one-third of an orbit since it was found in 1930). Astronomers quickly realized that there was a problem with Uranus, because it didn't seem to behave as calculated, wavering from the path expected if only the Sun and the known planets affected its motion. By the 1840s it was obvious that something was wrong, and two astronomers—Urbain Le Verrier in Paris, France, and John Couch Adams in Cambridge, England—independently predicted the mass and position of another planet beyond Uranus. The idea was that the gravitational tugs of this yet-unseen planet would explain the anomalous orbit of Uranus. While British astronomers dithered, Johann Galle and Heinrich d'Arrest, using the Frenchman's predicted positions for the new planet, spotted Neptune from Berlin in September 1846.

This provoked uproar in Britain, as claims were made for parity between Adams and Le Verrier in terms of credit for the prediction. The brunt of the responsibility for letting the discovery slip away needed to be borne by the professionals at the Royal Greenwich Observatory and within the universities, particularly at Cambridge. If amateurs with good equipment, such as James South in Kensington, had been privy to Adams's prediction then perhaps British honor might have been saved and Neptune discovered from within its shores.

Stung by all this, various amateur astronomers leapt into action. One of them was William Lassell, who had an excellent private observatory situated near Liverpool, later removing to the clearer climes of Malta. Like William Herschel before him, Lassell was skilled at constructing large reflecting telescopes, and with his champion he quickly discovered Triton, the massive moon of Neptune. But Lassell went further. Before long he was claiming that a ring like that of Saturn accompanied this new planet. That ring seems to have been either a figment of Lassell's imagination or a spurious image produced by his homemade instrument. Eventually rings around Neptune *were* discovered, but only a couple of decades ago, and they are much too tenuous to bear any relation to Lassell's claim.

Again these rings were identified through the tracking of occultations. Astronomers in the 1980s witnessed dips in their traces of stellar intensity before and after the star passed behind the planet itself, as with Uranus, but in this case the dips were not symmetric about Neptune. A strong decrease on one side was not repeated on the other, and when a pair of dips *did* occur they were not equally distant from the planet. This left a bit of a quandary for the astronomers: had they identified Neptunian rings or not?

By this time a dark, thin ring about Jupiter had been spotted using the *Voyager* spacecraft, leaving Neptune the odd man out of the gas giants if it lacked a ring system. Thus the betting was on rings being confirmed when *Voyager 2* at last reached Neptune in 1989. Sure enough, those rings were found in accord with the occultation data, and the reason for the ambiguity became obvious: rather than having complete circular rings, the dust orbiting Neptune seems to be concentrated in short arcs, as in Figure 12-5. The occultation observers by chance had intersected some arcs, but not others, producing their puzzling results.

PROBING SATURN'S RINGS

Even in the case of Saturn, whose rings were discovered by Galileo in the early seventeenth century (he described them as horns or handles jutting out from the planetary disk), occultations can still tell us much about the structure of the debris circuiting that planet. The timing of the roller-coaster ride followed by the intensity of light from a carousing star allows far better resolution than we can obtain from direct images of the rings. The photographs of Saturn

FIGURE 12-5. The ring arcs of Neptune as imaged by *Voyager 2* in 1989. Neptune itself is at lower right, somewhat overexposed with the image contrast stretched to make the rings visible.

from the *Voyager* spacecraft encounters are wonderful, but our detailed knowledge of the ring structure derives from artificial occultation data obtained by recording the intensity blips of stars whose light was intercepted as the spacecraft swept by the rings. Because of the data collected in that way we know that, rather than being broad, flat, featureless bands, the rings of Saturn actually contain many thousands of individual strands, their dynamical behavior affected by the gravitational tugs of its several dozen moons.

OCCULTATION OMNIBUS

An eclipse of the third kind—an *occultation*—occurs when a Solar System body traverses our line of sight to some distant cosmic light source. Studies of occultations allow both the nature of the light source (single star, double or binary star, galaxy, quasar) and that of the occulting object (lunar limb, asteroid size and shape, comet, planetary atmosphere, planetary ring) to be investigated.

One can think of a form of eclipse of yet another class, though. What about when three Solar System objects line up? The three involved in solar and lunar eclipses are the Earth, Moon, and Sun, but other combinations are possible. For example, both Venus and Mercury have smaller orbits than that of our planetary domicile. Do they ever cross the face of the Sun?

13

...and a Fourth

The planets show again and again all the phenomena which God desired to be seen from the Earth.

Georg Joachim Rheticus (1514–1576)

Because Mercury and Venus are sunward of the Earth they, like the Moon during a solar eclipse, may pass across the face of the Sun. Such events do not occur frequently. Eclipses of the third kind are called occultations; the present subject, eclipses of the fourth kind, are termed *transits*.

Earlier we noted that, if the Moon orbited us in the same plane as the Earth itself orbits the Sun (the ecliptic), it would be inevitable that eclipses (both solar and lunar) would happen every month. Because the lunar orbit is tilted at five degrees to this plane, however, they occur with a lesser frequency.

The same reasoning applies to Mercury and Venus: likewise they do not orbit in the same plane as the Earth. Mercury's orbit is tilted by just over 7 degrees, and that of Venus by 3.4 degrees. Using those angles one can go through the same rigmarole as for the Moon, to derive ecliptic limits on the nodal longitudes producing a transit. We won't trouble to step through those calculations though. Let's simply note that, due to regularities in their orbital motion, transits of both Mercury and Venus occur in distinct cycles. Those cycles will be discussed later. Historically, the transits of Venus comprise the most significant and rare phenomenon, and so we will first discuss those, and then look at Mercury.

THE TRANSITS OF VENUS

Looking up "Venus" in a gazetteer of world place names, one would find there are towns with that name in Florida, Pennsylvania, and Texas, a Venus Bay near Melbourne in Australia, and a Point Venus (actually *Pointe Vénus* in the local French) in Tahiti. It was from there that James Cook and his companions observed the transit of Venus in 1769. Why they traveled so far and at such expense to witness this celestial event is a matter of some significance. As we will see, Cook's expedition had enormous ramifications for the European settlement of the Pacific.

As it passes across the face of the Sun, Venus appears about one-sixtieth of a degree wide, equivalent to one part in 30 of the solar diameter. This means that it *could* be observed with the naked eye using a suitable filter, but it is better and safer to use a telescope projecting an image onto a screen. Venus then would look like a circular sunspot taking typically six hours to cross the face of the Sun, depending on which path across the disk (which *chord*) it follows. No living person has seen such a thing, because none has occurred since 1882.

In principle Venus could have been seen in transit before the invention of the telescope early in the seventeenth century, but no such observation prior to 1600 has been identified. This is hardly surprising since a transit is such an infrequent event. At about the same time as the first telescopes were being turned to the skies by Galileo and his followers, Kepler was evincing the laws of planetary motion. These laws enabled him, in 1629, to predict that transits of both Mercury and Venus would occur in 1631. It happened that he died in 1630. Even if he had lived, Kepler knew he would not see the Venusian transit, as it was only visible from much further west than Europe, from the Americas and the Pacific. But his prophecy of its occurrence was in itself a triumph. Most European astronomers had no doubt that Kepler was correct about that, despite the lack of visual confirmation, because the predicted transit of Mercury *was* seen, from Paris in particular, in November 1631.

Kepler did get something wrong though. He thought there would be no more Venusian transits until 1761, whereas in England Jeremiah Horrocks realized that Kepler was mistaken, just in time for the 1639 transit. In principle this event was visible over a wide area, but Horrocks only managed to alert one other observer to his calculations. Between clouds and between church services— it was a Sunday and he was the curate at a small village just north of Liverpool—by projecting an image onto a screen using a small telescope Horrocks glimpsed Venus creeping over the face of the Sun.

> O most gratifying spectacle! The object of so many earnest wishes, I perceived a new spot of unusual magnitude, and of a perfectly round form, that had just wholly entered upon the left limb of the Sun, so that the margin of the Sun and spot coincided with each other, forming the angle of contact.

Horrocks was able to monitor the transit for only half an hour before sunset, but his observation of the planet starkly and sedately moving over the disk of the Sun was confirmed by his friend William Crabtree, who lived about 30 miles away, near Manchester.

We saw earlier that total solar eclipse tracks crossed Britain in 1715 and 1724, followed by a hiatus of two centuries. This was just due to chance, in essence. Were the two transits of Venus in 1631 and 1639, followed by a gap of over a century, similar chance occurrences?

The answer is no. Transits of Venus occur as regular as clockwork, following a simple cycle. The transits always occur in pairs separated by 8 years. The Venusian orbit lasts for 8 parts in 13 of a year, an example of a resonance (the technical term is a "commensurability") in the Solar System. This means that after eight of our orbits Venus has circuited the Sun 13 times, and returns to more or less the same position relative to us. Due to the precessional movements of both planets the alignment does not repeat precisely. If in one nodal passage Venus happens to be near conjunction, resulting in a transit, eight years later it has shifted such that its apparent path has moved, but it is still within the ecliptic limit and a transit recurs, following a different chord across the Sun. Another eight years later the node has moved beyond the ecliptic limit, and no transit can take place. There is then another century or so of nodal movement before an alignment can occur again.

The clockwork of the heavens is such that transits of Venus occur with separations of 8.0, 121.5, 8.0, and then 105.5 years. Two transits occur spaced by 8 years; then there is a 121.5-year gap before there is another pair at a time of year 6 months away from the first pair; then another 105.5-year gap producing a pair again in the original month. This is because the nodes of the orbit of Venus pass across the Sun in early June at the descending node, and early December at the ascending node. (Note that taking 105.5 away from 121.5 you get 16, which is twice 8, showing the clockwork in action again.)

Including 1639, only five transits of Venus have ever been observed, in December of that year by Horrocks and Crabtree, in June of 1761 and 1769, and in December of 1874 and 1882. None occurred during the twentieth century. Without too much mental exhaustion you should be able to see that we are due to be treated to a repeat performance soon: the first transit of Venus for more than 120 years is scheduled for June 8, 2004.

This forthcoming transit is centered on about 08:20, Universal Time (UT: the standard time for the prime meridian passing through the Greenwich Observatory in London, England). If you plan to be in London on that day, the time on your watch would be an hour later, because Britain will be using summer time (clocks moved forward an hour) in June. To save confusion I will use UT for all times here.

The transit begins when Venus first appears to make contact with the solar disk at 05:15, about an hour after sunrise in Britain, and continues until 11:28, so that the event lasts for more than six hours in all. In Continental Europe, and further to the east, the Sun will have risen earlier and consequently be higher in the sky. One could argue that the optimum location from which to view the transit would be where the Sun is close to overhead at mid-transit, and that would indicate somewhere in the Middle East, such as Saudi Arabia. There is also a higher chance of the sky being clear there than in London. Much further east, such as in Japan, only the onset will be visible, the Sun setting before the transit ends.

For American viewers, the advice must be to head east. If you are enthusiastic (and wealthy) enough, head for Europe or beyond. If you stay in North America, you need to be close to the Atlantic seaboard. For example, Venus will be near mid-transit when the

Sun rises as seen from Boston. As far west as the Mississippi the end of the show will be visible, as the planet slips off the Sun's face soon after it rises over the eastern horizon.

But what if it's cloudy? At least there is not another century to wait. On June 6, 2012 another transit of Venus will occur. This time you would need to travel to eastern Asia or Australia to get the best view. After that, Venus does not align again with the Sun until December 11, 2117 and December 8, 2125.

THE TRANSITS OF MERCURY

Given that Mercury is smaller than Venus, more distant from us, and also inclined at a greater angle to the ecliptic plane, you might guess that transits of Mercury occur less frequently even than the rare Venusian transits. But you would be wrong. Mercury crosses the face of the Sun 13 times a century on average.

This does not imply, though, that Mercurial transits are spaced by even gaps of 7.7 years. Like Venus, Mercury follows a cycle with steps of certain length, quantized as multiples of an Earth year, but those steps are uneven. For Venus the steps are a regular sequence of 8, 121.5, 8, 105.5 years, but for Mercury there are interleaved cycles of 7, 13, and 33 years. The outcome is that Mercury's transits may be separated by only 3 years, but there may be up to a 13-year gap.

As for Venus, the dates of Mercurial transits are spaced by six months: they all fall within a few days of May 8 and November 10. Those dates define a position of the Earth in its orbit, and if on either date Mercury happens to be near its appropriate node (descending in May, ascending in November) then a transit will occur.

Another regularity is also produced. In a November transit, Mercury is near its perihelion, making it more distant from Earth, and so its disk appears small: only about one part in 190 of the solar diameter. Conversely, a May transit happens while Mercury is near aphelion, making it appear larger, about one part in 160 of the solar disk. (Recall that Venus appears to be about one part in 30 the solar diameter when in transit, so that Mercury always represents a rather smaller spot passing over the Sun.) This behavior makes May transits slightly easier to follow, but they occur only about half as often as November transits. This is because at aphelion the planet is moving slowest, and consequently it is less likely to pass across the Sun during the critical window. November transits independently follow a cycle with 7-, 13-, and 33-year intervals, while May transits are governed only by 13- and 33-year gaps.

Recent and upcoming transits of Mercury are as follows.

1970: May 9
1973: November 10
1986: November 13
1993: November 6
1999: November 15
2003: May 7
2006: November 8
2016: May 9
2019: November 11

There is then a 13-year wait until 2032 for the next opportunity.

Transits of Mercury typically last several hours, the longest in recent times being the 7 hour 47 minute behemoth of 1878. That which occurred in November 1999 was much briefer, lasting for but 50 minutes. This was barely a transit at all because, depending upon the viewing latitude, Mercury only just managed to break onto the Sun. This is called a *graze*, a rather rare event. Observers far enough north saw Mercury enter the face of the Sun in its entirety, but not venture far from the edge before terminating its fleeting visit (see Figure 13-1), whereas those further south saw the planet simply skim along the solar limb.

A transit is something well worth seeing at least once in your life, and there is a better window of opportunity in May 2003, when all longitudes from Europe east across Asia to Japan are favored as Mercury traverses the face of the Sun in a much deeper fashion.

A small telescope projecting an image onto a screen is what is needed, or a proper filter fitted to a telescope allowing direct view-

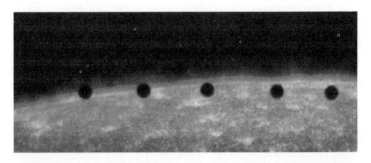

FIGURE 13-1. The transit of Mercury over the edge of the Sun on No-vember 15, 1999, as recorded with an ultraviolet telescope on board a satellite called *TRACE (Transition Region And Coronal Explorer)*. The five dark spots show the movement of Mercury over a time-span of almost 30 minutes.

ing. Mention was made much earlier of the ubiquitous Hα filter used in solar observing. Such a filter is especially useful in this case because it dims the brightness of the solar disk while making the chromosphere and corona visible, because it permits the transmission of only a single red light wavelength emitted by hydrogen. As a result Mercury (or Venus, if you watch in 2004 and 2012) may be seen silhouetted against the chromosphere before and after it meets the solar limb, whereas a simple gray (neutral-density) filter leaves the chromosphere virtually invisible.

Let us leave Mercury with a historical note. The first recorded transit was seen from Paris in November 1631. Pierre Gassendi was able to watch Mercury cross the Sun's face after receiving Kepler's prediction, confirming that the calculations were correct. A few other European astronomers who had heard of Kepler's work did likewise. The transit of Venus in the following month was unseen due to geographical considerations: there was not the time for observers to travel to the Pacific Ocean from where it could have been seen. We described above how Horrocks watched the Venusian transit in 1639, based upon his own calculations and ignoring the slip made by Kepler. For some reason the transit of Mercury in November 1644 passed unnoticed.

Another Englishman, Jeremiah Shakerley, later computed that a transit of Mercury would occur in 1651, but found it would be night in Britain when it occurred. Accordingly he traveled all the way to Surat in India to observe it. Solar eclipse chasing became a major pursuit in the Victorian era, but perhaps we should accord Shakerley some recognition as the first individual to make an intercontinental voyage in the quest for a glimpse of an eclipse of the fourth kind.

INGRESS AND EGRESS

Ingress and *egress* are the terms usually employed for the phases when Mercury or Venus are entering and leaving, respectively, the solar disk. Such terminology may also be used for eclipses and occultations, along with their synonyms *immersion* (or entrance) and *emersion* (or emergence).

The different contact points for a transit are shown in Figure 13-2. Ingress lasts from when the planet meets the solar limb (contact I) until the instant at which the planetary disk is totally encompassed (contact II), and similarly for contacts III and IV at egress. These junctures are analogous to the contacts occurring in an annular solar eclipse, except that now the dark object is much smaller than the Moon. Without a suitable filter one cannot properly observe contacts I and IV, making accurate timings difficult, and astronomers try to time instead contacts II and III, but fixing those instants is not easy either.

This difficulty is caused by a phenomenon called the *black-drop effect*. As the planet is completing its ingress, instead of a simple dark disk its image seems to be distorted into the form of a raindrop, as if a thread or ligament of material has attached it to the solar limb, pulling it out of shape. The appearance of Venus in 1769 is sketched in Figure 13-3. Contact II is strictly when that thread seems to break, with a circular silhouette being formed, completely surrounded by the Sun. Similarly at egress, for contact III, the time in question is just before the thread appears.

In visual observations the eye is often deceived. Apart from the black-drop effect, observers of the transits of Venus have reported the planet to appear surrounded by a luminous patch or aureole (Figure 13-4), sometimes with a bright spot on the dark

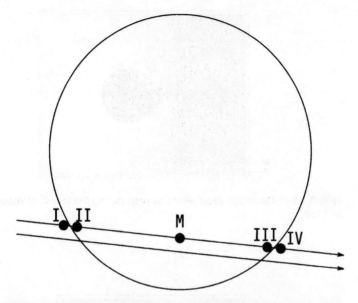

FIGURE 13-2. The contact points in a transit are labeled with Roman numerals from I to IV. Here M is the mid-point. The chords followed by Venus in transit across the face of the Sun during a transit depend especially upon the latitude of the observer. By measuring the times of ingress and egress accurately, the precise chord taken can be determined, and with two timings/two chords (as depicted here) from observers at known locations, it is feasible to calculate the distance to the Sun with some accuracy.

disk. These optical effects, due to scattering by the atmosphere of Venus, were unsuspected until transits were first watched. They limit the accuracy with which the phenomena may be timed, and that has important scientific and practical repercussions, as we will now see.

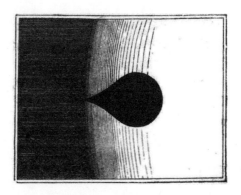

FIGURE 13-3. The black-drop effect as seen during the transit of Venus in 1769.

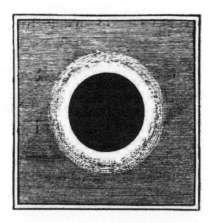

FIGURE 13-4. During the 1769 transit a bright ring or aureole was seen around Venus, caused by its atmosphere.

THE SCIENTIFIC USE OF TRANSITS

Leaving aside the transits of Mercury in 1661, 1664, and 1674, the next important occurrence in this field of endeavor was in November 1677 when another was observed by Edmond Halley, who at the time was on the island of St Helena, engaged in a survey of the South Atlantic.

This episode is significant because Halley later expounded a technique for using complete transit timings to determine the distance between the Earth and the Sun (that is, the astronomical unit or AU), a measurement that was sorely wanted. It was contemporary ignorance of the scale of the Solar System that led to the inaccuracy of Halley's computed track for the 1715 eclipse (see Chapter 7). An even more important consideration was that navigational accuracy at sea required precise knowledge of the future positions of the Sun, Moon, and planets.

Many authors credit Halley with inventing the transit technique. Actually it had been a Scottish mathematician, James Gregory, who first propounded the idea, in 1663, but it was Halley's later description of the concept that led to attempts to put it into action. He read a paper on this topic to the Royal Society in 1691, but did not publish his analysis until 1716. Halley realized that only transits of Venus, not Mercury, would afford a feasible avenue for determining a better value for the mean Earth–Sun distance. Knowing that those transits would not occur until 1761 and 1769, Halley recognized he would not live to put the technique to the test (he died in 1742). Nevertheless, he was happy to leave his reputation to posterity, just as he knew that he would not be around to see the return of the comet that bears his name, in 1758.

The significance of these transits we should put in the context of the era. The quest for a practical method for determining the longitude of a ship at sea was an overbearing desideratum at the time, as we discussed in Chapter 3. Astronomers were engaged in a race to develop some accurate method that would allow the position of a vessel far from the sight of land to be easily and accurately measured. Reckoning the distance to the Sun was not some abstract piece of scientific curiosity: it bore the promise of more accurate celestial tables and thus improved navigation. In consequence the British (and other) governments were strongly interested in having the transits utilized by their astronomers to ascertain that quantity, from which the distances and motions of the Moon and planets could be computed using Kepler's Laws. This explains the expense and effort put into this endeavor, as exemplified by the voyage of Lieutenant James Cook detailed in the next section.

THE FIRST VOYAGE OF LIEUTENANT JAMES COOK

The explorations of James Cook in the Pacific are well known, but the primary purpose of his first voyage, from 1768 to 1771, is not so extensively recognized. That purpose was to observe the anticipated transit of Venus on June 3, 1769, from Tahiti. In fact there had been a transit in 1761, as noted earlier, and the peculiar tale of that event will be told a little later in the book. For the time being, we pick up the story with Cook having been dispatched to the South Seas on board a small ship named the *Endeavour*, loaded with men and supplies, but most especially carrying various telescopes, a pendulum clock, English astronomer Charles Green, and Swedish naturalist Dr. Daniel Solander. Also on board was Sir Joseph Banks, who left a long-term mark on British science, serving as President of the Royal Society for several decades after his return.

Overcoming mishaps along the way, Cook and his party arrived at Tahiti a couple of months before the transit was due. They needed plenty of time to set up their temporary observatory. Nowadays the location may be denoted Pointe Vénus on a map, one of the few non-Polynesian names thereabouts, but Cook called it Fort Venus, as he had to guard his establishment to stop the indigents from stealing the equipment and supplies. Similarly the island appears in the records of the expedition not as Tahiti, but as *King George's Island*, for the sovereign. A British party had charted it only a year or so before, just in time for planning Cook's expedition, a good base in the largely unexplored South Pacific being required for the transit timings.

The astronomical observations of the party went well, although with some drawbacks, the significance of which we will come to later. To quote from Cook's personal journal: "Saturday 3rd June. This day prov'd as favourable to our purpose as we could wish, not a Clowd was to be seen the whole day and the Air was perfectly clear, so that we had every advantage we could desire in Observing the whole of the passage of the Planet Venus over the Suns disk: we very distinctly saw an Atmosphere or dusky shade round the body of the Planet which very much disturbed the times of the Contacts particularly the two internal ones." This brings us to the secondary aim of the voyage. Cook took with him a sealed envelope, containing his instructions for the rest of his mission. Although the gist of it seems to have been common knowledge in England, the contingent on board the *Endeavour* could not have known for sure what those orders were until after the transit. By that time they were already nine months into a voyage that was to last for almost three years.

The secondary task is now well known: explore the southern oceans to search for the hypothesized great southern continent, *Terra Australis Incognita*, and claim it for the British crown. There is no great landmass in the far southern Pacific, so they sailed that ocean in vain. However, Cook charted and claimed New Zealand and the east coast of Australia, even though the French and especially the Dutch had been there before, with the result that those became British colonies. French Polynesia, from where the transit was observed, is obviously a different story.

Earlier I challenged you to look up "Venus" in a gazetteer, identifying Pointe Vénus in that way. If you do the same thing for "Mercury," you will find a small town in Nevada, 60 miles northwest of Las Vegas—perhaps because a thermometer's mercury soars in the desert—and also a settlement known as Mercury Bay in New Zealand. This is on the Coromandel Peninsula, east of Auckland, and to the northeast in the Pacific is Great Mercury Island. How did these names come about?

It happens that, like 1631, 1769 was a double transit year. After the one by Venus in June, Cook knew that Mercury would also pass over the face of the Sun in November. This event he planned to put to a different purpose. Concurrent lunar eclipse observations from separate locations allow their difference in longitude to be derived, by comparing the eclipse timings to the local solar time. This method had been used, for example, to determine the distance west from London to the Caribbean and the Americas, certain eclipses being visible both from there and back in Europe. Cook, however, was right around the other side of the planet, meaning that he could not watch an eclipse at the same instant as astronomers viewed it from England.

The instant of the transit of Mercury had been precalculated with some precision. It provided a natural clock in the sky. By comparing the time at which it was observed with the local time according to the Sun, Cook could determine the longitude of New Zealand. For this reason he scouted the North Island, eventually choosing the place now called Mercury Bay because the Maoris there seemed less hostile than elsewhere.

The rival method to astronomical observations for determining longitude at sea was John Harrison's chronometer, which eventually triumphed. On this first voyage Cook had no marine clock, so he needed to rely on astronomy for determining time and longitude. The transit of Mercury in late 1769 provided a particular opportunity. At other places he used both the lunar distance technique, and the eclipses of the four Galilean moons of Jupiter, methods mentioned in Chapter 3. On his subsequent voyages of discovery Cook had an excellent chronometer on board, but still made astronomical observations to verify that clock's accuracy.

Cook eventually arrived back in England in 1771, after many tribulations, with his much-awaited transit timings. The astronomer Green did not make it, having died on ship. Our next port of call is the usage that was made of the transit observations, but to understand that we must step back to 1761.

PARALLAX AND DISTANCE OF THE SUN

After Halley's death, others took up the cudgels in persuading the British government that the 1761 and 1769 transits of Venus represented an opportunity that should be seized, confident that the problems of maritime navigation would be solved once the distance to the Sun was known with sufficient accuracy.

This distance was largely a matter of conjecture. Halley himself had decreased his estimate for the astronomical unit by a factor of four, but even his final value was about 30 percent too high. On the Continent, astronomers had tried using observations of Mars to measure this parameter, but their results did not agree.

The basis of the method used was parallax. This is easy to demonstrate. Hold your arm out straight, with one finger pointing upwards and some far-flung background beyond it. Repetitively looking at the background with one eye shut and then the other, your finger seems to jump left and right. If you measure the distance it appears to move, and also the separation of your eyes, then you could determine the length between your head and your finger. A tape measure may be a simpler way, but one cannot stretch a tape measure to Mars. In the astronomical context Mars is equivalent to the finger, and the far panoply of stars is the unmoving background. So what is the analogue of the separation between your eyes, providing the parallax effect? In 1751 French astronomers observed the apparent position of Mars against the stars both from Paris and the Cape of Good Hope (South Africa), providing a baseline of over six thousand miles. From their derived distance to Mars they calculated the astronomical unit, but their value was too high.

Remember these were the days before photography. Unlike in Eddington's 1919 eclipse expedition, it was not possible to photograph Mars surrounded by a star field at a particular time, bringing the plates home for close comparison. A more accurate visual method was needed, and a transit of Venus afforded just that. The basis of the technique was as follows.

Just as your finger moves as you blink eyes, if observers are well separated they will see Venus take different paths or chords across the Sun during a transit. That's parallax, and the idea in this context is represented in Figure 13-2. If the distance between those chords is determined, then it is feasible to compute the distance to Venus, and to the Sun. The trick is in measuring the locations of the chords with sufficient precision. This could not be done *directly* because of the uncertainty of the azimuthal positions on the Sun's face of the contact points, the finite size of the disk of Venus, and so on. But there was a refinement, as follows, making the path determination possible.

The rate at which Venus appears to move across the sky during a transit may be calculated quite accurately on a theoretical basis. If the transit is timed, then the angle Venus moved through during that time interval may be calculated, and from that the chord taken across the solar disk determined with some precision. It was anticipated that, if the contact junctures were timed to within a few seconds, then with a transit lasting for five or six hours the chord would be extremely well defined. With two observers separated by some known distance (that is, if their latitudes and longitudes were known), it would be feasible to arrive at the solar distance with an accuracy far superior to all previous measures.

The simple idea that Cook and his party were sent to Tahiti to measure the distance to the Sun is a little misleading, however. From the above description, it is clear that *two* observation points are required, separated by as far as possible. Cook's expedition provided just *one* of them, so it is incorrect to think that the Tahiti measures could be used on a stand-alone basis. We will come back to this later.

THE TRANSIT OF 1761

Going back a little before Cook's voyage in the late 1760s, let us consider what happened with the earlier transit, that in 1761. As this was approaching, astronomers were not inactive. It was known that the event would be visible in its entirety in a band stretching from northern Europe across the landmass to the southeastern parts of Asia. To each side of that band either only ingress or only egress could be observed, but that did not preclude useful timings being obtained: an ingress observed at one point could, at least in principle, be combined with an egress timing elsewhere, so long as their geographical coordinates were well determined.

In London the Royal Society organized several expeditions. The fifth Astronomer Royal, Nevil Maskelyne, set sail for St. Helena. From there and at the Cape of Good Hope the egress would be visible. Far better were locations from which both ingress and egress could be seen. The French, British, and several other nations set up a number of temporary observatories in suitable places. The shortest transit time was observed from Tobolsk, 300 miles east of the Ural Mountains in western Siberia. In India the duration was three minutes longer. It follows that, although the transit lasted for many hours, to differentiate between the chords required timings good to a matter of seconds.

There is another, related, consideration. To get the best parallax, the widest possible latitudinal separation is needed. In the event the furthest northern point at which the entire transit was timed was Torniö (on the current border between Sweden and Finland), at a latitude near 66 degrees; the furthest south was Calcutta, at 22.5 degrees north. The separation of these is not much more than 40 degrees. Other sites (such as South Africa) were much further south, but only the egress was observable from there. On the other hand, the longitude coverage was excellent, from Jesuits in Beijing to astronomers sent to Newfoundland, 120 observers in all, but with the restricted latitude range covered, they were all watching similar paths across the Sun.

This meant that the data collection was far from optimal, and it was realized that the much hoped-for improvement in astronomical and navigational knowledge would not result, at least immediately. The transit *could* have been observed from start to finish on the equator in Indonesia, and even further south (the Dutch had already literally run their ships into New Holland, now called Australia). Why wasn't it?

DIXIELAND BLUES

All will be familiar with the southern parts of the United States— the old Confederacy—being termed Dixieland or simply Dixie. There are several theories concerning the origin of this moniker. A leading idea is that it derives from the name of an English astronomer, Jeremiah Dixon. With his compatriot, Charles Mason, Dixon surveyed the border between Maryland and Pennsylvania from 1763 to 1767, defining the famous Mason–Dixon line. Until the Civil War in the 1860s this was considered to be the demarcation between the free states and the areas of black slavery below.

In 1760, however, the pair was looking at heading east towards Asia rather than west towards the Americas. The British wanted to send a transit observing team to Bengkulu in Sumatra, four degrees south of the equator. In those days nearby Jakarta was named Batavia, the capital of the Dutch East Indies. Mason and Dixon were engaged for the task, and in March 1760 they set sail from Portsmouth on a Royal Navy ship. Before slipping out of the Channel, their vessel was attacked by a French frigate. Eleven sailors were killed, and 37 wounded, the British limping back into Plymouth.

Not surprisingly Mason and Dixon had lost much of their enthusiasm for the adventure, and despite the navy offering to provide a mighty escort out of the Channel after their vessel was repaired, they wrote to the Royal Society petitioning for their destination to be switched to the Black Sea. One might suggest that this would involve an even more dangerous voyage through the Mediterranean, but it seems that it was not only the French guns that worried Mason and Dixon. On their abbreviated venture the landlubbers had been stricken with seasickness, and they felt they could not stomach a voyage down through the Atlantic and across the Indian Ocean. This they were charged to do, though, in a forceful rejoinder from London. Nevertheless, a postscript to their instructions allowed them some discretion, and in the event they decided to halt at the Cape, from where they observed the egress. This was just as well: in the interim the French had seized Bengkulu, so that Mason and Dixon would hardly have been afforded a welcome there.

The timings our heroes made at the Cape were useful in the analysis of the transit carried out by mathematician James Short back in London, but there was a problem. A French expedition had gone to Rodrigues, a little island just east of Mauritius, where astronomer Alexandre Pingré watched the egress. Jacques Cassini, Director of the Paris Observatory like his father, grandfather, and great-grandfather before him, supplied that egress timing to Short. (Just because their countries are perpetually at war does not mean that scientists will not collaborate.)

Unfortunately the value obtained from the Rodrigues recording was discrepant when compared to the relatively nearby Cape, and Short thought that Mason and Dixon, who were already subject to some opprobrium, had mistimed the event by precisely one minute. By making this "correction" Short derived a distance to the Sun that was more than 10 percent lower than the real length. In fact the error was due to the longitude quoted for Rodrigues being out by a quarter of a degree. In the late nineteenth century the American astronomer Simon Newcomb, with the advantage of valid geographical coordinates for the observation sites, reanalyzed all the 1761 transit timings and showed that they were consistent with the true solar distance, which by then had been determined by other means.

Back in the 1760s this was not known. It seemed that the transit of Venus in 1761 had passed by without the necessary timings having been made with a sufficiently wide geographical spread. There was a determination that the opportunity in 1769 would not be similarly wasted.

PREPARATIONS FOR THE 1769 TRANSIT

Both the British and the French redoubled their efforts in the quest to obtain the desired benefit from the 1769 transit. The middle of the event was at about 22:20 UT (i.e., London time), so it was clear that stations in the Pacific were required if the entire six-hour transit was to be followed.

Simplistically, one could imagine that locations further east (for instance in the Caribbean) might be able to see the ingress, those further west (in India) the egress, and only at longitudes for which local midday is near 22:20 UT would the complete transit be observable. From that perspective Tahiti was an excellent prospect. Hawaii, which Cook was to map on a later voyage and name the Sandwich Islands, meeting his death there in 1779, would have been at a good longitude, although a more southern latitude was desired.

But that discussion *is* too simple. The transit was on June 3, only 19 days before the summer solstice, and so the Northern Hemisphere was tilted toward the Sun. In consequence the event might be seen throughout the Arctic, almost independent of longitude. Lapland is often called the Land of the Midnight Sun for a good reason, and it was realized that the transit could be observed from, say, the Russian town of Murmansk, and right across northern Siberia to the Pacific and thence Canada. At that time of the year the Sun is above the horizon for most of the day at such latitudes.

This meant the observational baseline could be stretched far to the north, and an international effort was organized, the British taking responsibility for extrapolating that baseline as far south as possible. Apart from Murmansk, and Hudson's Bay, ingress-to-egress timings were made in Norway. In fact it was our old friend Jeremiah Dixon who went to Norway, this time unaccompanied by Charles Mason. The idea that Cook was sent to Tahiti because the transit could not be observed from anywhere near the longitude of Britain is therefore incorrect: paradoxically, he sailed to Cape Horn and then westwards for reasons of *latitude*. That is, Cook's party went to the Pacific in order to observe the transit from as far *south* as possible.

One other important location from which the transit in 1769 was observed deserves special mention, a French expedition. It was the Abbé Jean Chappe d'Autoroche who had watched the

1761 transit from Tobolsk, in Siberia, an observation mentioned earlier. This time he wanted to sail to the Solomon Islands. These islands lie in the western Pacific, northeast of Australia, and were the site of ferocious fighting during the Second World War. Back in the eighteenth century the Solomons were under Spanish control, but despite intending to take two Spanish naval officers along the court of Spain refused him leave, suspecting him of wanting to spy out the territory on behalf of France. Thus Chappe sailed across the Atlantic, through the Caribbean, and landed in Mexico at Veracruz. From there his party traveled overland through Mexico City at great personal danger from banditos, the local Viceroy then providing them with an escort of soldiers as they pushed on to the Pacific coast through Guadalajara. From San Blas they sailed, with some difficulty, northwest towards Cape San Lucas, the tip of Baja California, and observed the transit from San José del Cabo. This has been the source of much confusion, "San José, California" being a totally different place, deep in Silicon Valley.

Chappe got to this lesser-known San José a fortnight before the transit and fixed its latitude by observing the culmination of stars, its longitude using the moons of Jupiter, and then the transit. A complete success, except that a contagious disease—a strain of typhoid it seems—was already sweeping San José when his party arrived. Ignoring the danger, Chappe insisted on remaining not only for the transit, but also thereafter for a lunar eclipse on June 18. Timing of that eclipse was required to secure the site's longitude, an essential parameter if the transit project was to succeed. By then Chappe had himself succumbed to the illness. He died six weeks later, as did one of the Spanish officers, but the remnants of the party ensured that the invaluable timings were returned to Europe.

ANALYZING THE 1769 RESULTS

Back in England, the task of analyzing the available timings fell to Thomas Hornsby, the Savilian Professor of Astronomy at Oxford. His selection from the data available from Tahiti was peculiar, however. The observations from Cook himself, Green and Solander, all with their own telescopes, showed a scatter of ten seconds or more in some of the contact timings, as foreshadowed by the quote from Cook's journal given earlier. Hornsby seems to have selected the values from Tahiti that fitted in best with what he expected, based upon the calculations he had already completed using information over the shorter baselines. That is, because Cook did not arrive back until 1771, Hornsby had already made calculations using combinations of readings from Wardhus in Norway, Murmansk in Russia, and Hudson's Bay in Canada. Later came the timings from the French at San José del Cabo in Baja California, and finally the data from Cook's party in Tahiti. It seems that by the time that he received the final set of timings, Hornsby had already made up his mind what the answer should be.

This selection of data is dubious in itself, but also there were other observations available, which Hornsby ignored. One wonders how his report would be treated if subject to the rigorous perusal typical for modern-day scientific papers. Perhaps not by chance, Hornsby's final value for the Earth–Sun distance was much the same as that he had derived using the 1761 transit.

The matter did not sit there, though. The timings from Rodrigues in 1761 were misleading because the longitude of that island was imprecisely known. Cook and colleagues in Tahiti in 1769 determined the longitude of Fort Venus in two ways: from the eclipses of the satellites of Jupiter, and lunar observations. Both results differed from the true longitude, measured later when marine chronometers were carried to Tahiti, by tens of seconds. Even if the observations of Cook, Green, and Solander had agreed with each other, still there was another inherent source of error making the final result for the Earth–Sun distance incorrect: the site coordinates were wrong.

To that extent, one has to say that the expeditions mounted to observe the transit of Venus in 1769 overall were a failure; a failure that cost many lives. Of course there were many spin-offs, such as those accruing from Cook's sealed-envelope orders (I am a citizen of Australia, and previously lived in New Zealand for some years), but basically the science did not work.

The transit observations from 1761 and 1769, so eagerly recommended by Halley and others, did not lead to improvements in navigational capabilities, but within a handful of years that motive anyway had been surpassed by other developments. As aforementioned, on his second and third voyages James Cook carried accurate marine chronometers modeled on Harrison's clocks and fixed his longitude using those.

MEASURING THE ASTRONOMICAL UNIT

The fundamental aim of the transit expeditions was to enable the astronomical unit—the AU, the Earth–Sun distance—to be determined. It would be remiss if I did not complete that part of the story.

It already has been mentioned how Simon Newcomb, in 1891, reexamined the 1761 transit data and, using the correct geographical coordinates for the observation sites, showed that the timings were consistent with the actual solar distance. In fact he did this likewise with the 1769 data, handling well over a hundred timings, some of which he had to reject as clearly erroneous.

Newcomb was not the first to attempt this reanalysis. Astronomers did not simply wait for the 1874 and 1882 transits to arrive. In the first half of the nineteenth century various attempts were made to exploit the 1761 and 1769 data. The German astronomer Johann Encke did this, but ended up with an answer making the solar distance somewhat larger than indicated by other techniques; the result was that the transit observations were distrusted until Newcomb demonstrated their veracity.

In both 1874 and 1882 renewed efforts were made to determine the AU through Venusian transits. In the former year the United States alone sent three expeditions to Siberia, Japan, and China to achieve northern sightings, and five groups to New Zealand, Australia, and Kerguelen Island in the Southern Hemisphere, the advent of photography allowing a permanent record of the phenomena to be made. The weather stymied much of the photography, and comparatively little success was met by the Americans or the numerous British, French, and Russian groups, and others. The Germans did better, obtaining clear weather at all six of the sites they had chosen. In 1882 a similar array of astronomers observed the path taken by Venus across the Sun, although American astronomers did not need to venture too far: the whole transit was visible from the eastern two-thirds of North America and all of South America.

Science moves on, though. In 1898 the large Earth-approaching asteroid 433 Eros was discovered. Within a couple of years, astronomers were using parallax observations of Eros in the same way as Mars had been employed earlier. Eros comes much closer to us than Mars, leading to a more accurate evaluation of the AU.

The invention of radar led to the ultimate determination of the AU. Again, Venus has been involved, although in a quite different way. By bouncing radio pulses off that planet and timing the echoes' return, the solar distance now has been measured with a precision unimaginable to Halley, Cook, and all others involved when transits of Venus were considered by many to be the only viable avenue to improved navigation.

PLANET–PLANET ECLIPSES

For the sake of completeness, there are a couple of other phenomena we might tidy up in our survey of peculiar types of eclipse. The first is trivial. In the Space Age a host of artificial satellites has joined our natural satellite, the Moon, in orbit about the Earth. These are eclipsed frequently. The time to watch for satellites is soon before dawn or just after dusk (because during the deep night, satellites in low orbits are within the terrestrial shadow, in eclipse). Far enough up that the Sun is still catching them, satellites in low orbits such as the space shuttle, the space station, or the Hubble Space Telescope typically take 90 minutes to circuit the planet. Those are only a few hundred miles up, higher paths taking longer to complete an orbit. The time to move from horizon to horizon typically is only a few minutes, but often one will see a satellite abruptly disappear, as it enters the shadow zone.

Devotees of satellite spotting also enjoy solar eclipses. In that situation the name of the game is predicting when a particular satellite visible in daytime (usually with binoculars) is going to pass into the shadow of the Moon—and then watch it actually happen. Catching artificial satellites being eclipsed, though, is a specialized modern-day sport. Let us return to natural events.

We have considered the Moon and planets crossing the Sun or the stars, Jupiter eclipsing the Galilean satellites, and measuring the sizes of asteroids and comets. Is it possible, though, that one planet could eclipse another? Venus, say, could cross the face of Jupiter, and because the former appears smaller than the latter this could be classed as a transit. Such an event *might* be seen around dawn or dusk if it happened that Venus were near maximum elongation (the greatest angular distance it achieves from the Sun) and Jupiter, on the opposite side of the Sun to the Earth, happened to line up. Alternatively Mercury might pass behind Venus and be occulted. Such things *must* happen—but not very often.

There is a thin line of differentiation between an occultation and a transit. One might say that a Galilean satellite is occulted when it passes behind Jupiter, but is in transit when it moves across the Jovian disk as seen from the Earth, the somewhat different direction to the Sun causing its shadow to be located elsewhere on that disk (see Figure 13-5). Both events might be thought of as forms of eclipse, which is why they merit mention.

The planets all orbit the Sun in the same direction, with orbital planes inclined slightly to the ecliptic. This prohibits planet-planet eclipses from occurring every year, but makes their occurrence more frequent than if they sped around the Sun with random orientations. Just how often *do* such events occur? As a long-term average, there are 7 or 8 years between solar transits of Mercury, and solar transits of Venus occur once every 60 years. The Sun covers a much larger target area than any of the planets, so one might anticipate that transits of one planet across the face of another would be rare birds indeed. This is indeed the case.

In 1591, while still a student at Tübingen in Germany, Johann Kepler ventured out into a cold January night with his teacher to

FIGURE 13-5. A transit of Io, one of the four Galilean moons of Jupiter, across the face of the planet results also in a form of eclipse, with its shadow being cast on the cloud tops below.

observe a predicted close conjunction between Mars and Jupiter. To their astonishment only one reddish spot could be seen in the sky, and they surmised that the two had aligned with each other. This would be the first planet–planet eclipse observed, except that precise backward computations show that Kepler's senses must have deceived him. What actually occurred is called an *appulse*,

Mars and Jupiter passing very close to each other, like a grazing occultation. Without a telescope, two decades before Galileo opened the heavens to closer inspection, the human eye was inadequate to differentiate the adjacent pair of tiny planetary disks.

In all recorded history there is only one definite observation of a planet–planet eclipse, and that was watched by but one man, from the Royal Greenwich Observatory in 1737. John Bevis was a physician from a country area a hundred miles west of London, who had done well for himself in the city, giving him the time and money to pursue his amateur scientific interests, including astronomy. Although not on the staff, he often observed the heavens from the great observatory, his expertise with the telescopes being well recognized.

One evening he was observing with a rather crude refractor (a lens telescope), with a focal length of 24 feet. In those days such simple telescopes tended to produce poor images with colored fringes around celestial objects. Through the long tube of this ungainly instrument Bevis saw a gibbous Mercury and narrow crescent Venus near each other in the sky, and rapidly closing. (Both planets display phases like the Moon; "gibbous" is the phase between when a half and a full disk is illuminated.) Clouds intervened, and it was eight minutes before Bevis could again espy the brilliant but slender Venus. The dimmer Mercury he no longer could detect. He surmised that Venus had eclipsed Mercury, but he was prohibited from seeing the smaller body emerge from behind the larger by yet more clouds, which blanketed the sky until the planets set in the west.

We met Urbain Le Verrier in the previous chapter, as one of the predictors of the existence of Neptune. Having been appointed Director of the Paris Observatory, in the mid-1800s he had drawn up tables of planetary positions and wanted to test their accuracy. Le Verrier seized upon Bevis's report from more than a century before as a stringent test. Sure enough he found the alignment would have occurred as described, with the slight refinement that Mercury was not completely covered by Venus. That part of Mercury protruding, though, was the dark part shadowed from the Sun, so Bevis could not have seen it in the glare of Venus. Bevis's northerly location was also critical; if he had been on the equator, the two planets would have swept past each other in an appulse.

A handful of other opportunities to witness planet-planet eclipses have been missed by astronomers over the past five centuries. In 1570 and 1818 Venus skimmed over Jupiter, but neither event seems to have been noticed. In 1705 an observer in Japan *could* have witnesssed Mercury practically touch Jupiter, as seen in the night sky, but it seems that none did.

Looking into the future, in 2037 Mercury will pass very close by Saturn, but not quite transit its disk or rings. If you choose your location carefully (go north, young men and women), you may see Mercury blotting out Neptune in 2067, but you'll need a decent telescope. After that, there is a transit of Venus over Jupiter in 2123, and in 2223 Mars will do likewise. Planet-planet eclipses, then, do not occur often.

LE VERRIER'S PLANETS

Having reintroduced Urbain Le Verrier above, we will now describe how he enters into transit observations.

As was mentioned earlier, after its discovery by William Herschel in 1781 the British wanted to name Uranus for King George III. Almost in retaliation, when Neptune was found in

1846 utilizing a Frenchman's predictions, his countrymen wanted to call it "Le Verrier" (I will leave it to the reader to consider why they did not instead try to honor their own royalty). The mythological name Neptune eventually prevailed.

The predictions of Le Verrier and Adams (see Chapter 12) were based upon the slight deviations between the theoretical and observed positions of Uranus. Some unknown body seemed to be tugging the planet along, and that was Neptune. A decade or so later Le Verrier had turned his attention to Mercury. The innermost planet also had an orbit that could not be explained using Newtonian gravitational theory coupled with the positions and masses of the known planets. It seemed that Mercury's perihelion point was precessing faster than expected (see the Appendix for an explanation of what is meant by "precession").

What Le Verrier suggested in 1859 was that several previously unsuspected small planets existed, orbiting closer to the Sun than Mercury. He reasoned that there must be several of these, rendering a total mass about one-tenth that of the Earth, and they had hitherto escaped detection because they were small and dim, compared to the bright solar glare. Le Verrier suggested that these unseen bodies were tugging Mercury along a little, explaining its slightly anomalous motion. So, how could they be discovered? The answer is simply to look at the Sun. Every so often one should appear in transit, taking minutes or hours to cross, not the 10 or 12 days of a sunspot.

This concept was greeted with enthusiasm and, sure enough, announcements of small dark spots transiting the Sun soon flooded in. The first to claim to have seen one was another Frenchman, Edmond Modeste Lescarbault. He was a country physician keen on astronomy, and he spent his leisure hours in a quest for these hypothetical tiny planets. Le Verrier was quickly acclaimed as the predictor of not only the outermost planet, but the innermost too. It was even given a name: Vulcan (aficionados of *Star Trek* take note).

The problem was that none of the putative observations could be verified, and the reports were inconsistent. In the heydays of visual astronomy, many claimed discoveries were figments of the fond imaginations of the observers involved. What was required was an opportunity for many observers to peruse a target at the same time. A total solar eclipse affords just such an opportunity: if Le Verrier was correct and there were one or more intramercurial planets, then they should be detectable during an eclipse, when the bright sunlight is largely blocked out.

The search for Vulcan and its putative companions became one of the major aims of the eclipse expeditions to the western United States in 1878 (as previously mentioned in Chapter 9). Observers in Colorado and Wyoming announced they had found not just one planet, but two, close to the Sun. But there were discrepancies in what was reported, and a major public argument ensued. The claimed discoveries were to the southwest of the Sun, whereas any body causing the charted perturbations of Mercury would need to have been to the east. The measurements from the two sites were inconsistent with each other, making some think that *four* new planets had been found. In the end it was realized that two rather faint stars in Cancer were all that had been detected, and so there were red faces all round.

Over the following years more rigorous scouring of the space around the Sun was conducted, when total solar eclipses allowed. These efforts did not go totally without reward: during the 1882 eclipse over Egypt, a comet was found that had previously defied discovery. A similar discovery, of a different comet, had been made way back in A.D. 418; and in 1948 observers in Kenya also found a comet in the eclipse-darkened sky. But no small planets within the orbit of Mercury, such as the Vulcan envisioned by Le Verrier, have ever made their existence known.

This is not surprising, because Le Verrier's analysis was based on a false premise, that the Newtonian gravitational theory is an adequate description of the laws of physics. When you are as close to the Sun as Mercury, it happens that Newton's theory breaks down. The explanation of Mercury's anomalous precession awaited Einstein's relativity theory, as was mentioned in Chapter 4. Along with the gravitational deviation of starlight, the explication of Mercury's orbital motion is one of the great demonstrations of the veracity of Einstein's theory.

There are, however, known asteroids that pass closer to the Sun than the Earth, making transits feasible. Since the first was found in 1932, several hundred Earth-crossing asteroids have been catalogued. Imagine that one was passing relatively close by our planet. If it happened to align with the direction of the Sun, we would see it transit the solar disk, taking between a few seconds and a minute to cross. (It would need to be close—within, say, a million miles of us—because all these asteroids are smaller than five miles in size, making them imperceptible against the solar disk if further away.) Very small dark spots quickly crossing the Sun have been reported many times, often by reputable and experienced observers, but the frequency of such events seems much higher than may be explained by the suspected flux of asteroids, leaving it all a bit of a mystery. As of yet no wholly intramercurial object has been found (nor indeed any intravenusian asteroid), but that does not mean that they do not exist. For example, the first asteroid always closer to the Sun than is the Earth was spotted only in 1998, and those still closer to the Sun would be even more difficult to find. We do know that all such bodies that may exist must be small, too small to cause any significant gravitational perturbations of the planets.

This is not the end of the story. After Neptune had been discovered and tracked for some decades, all the computations indicated that neither its path nor that of Uranus could be accommodated by the mutual gravitation of the known masses in the Solar System. This led Percival Lowell and others to think that there was another large planet still to be found. Lowell, who was introduced in Chapter 12, was a Bostonian who had made a fortune out of textiles, and so had money to spend on his favored hobby: astronomy. In Flagstaff, Arizona, he founded the great observatory that still bears his name. Apart from looking for evidence of life on Mars, the major task of the fledgling Lowell Observatory was a search for a further outer planet, and that project led to the discovery of Pluto in 1930. But the history is not quite as simple as that, as we'll see in the next chapter.

14

Stepping Beyond the Solar System

The astronomers said: "Give us matter, and a little motion, and we will construct the universe."

Ralph Waldo Emerson

We have been straying towards the fringes of the Solar System, and now we have just about reached the edge. From 1979 until 1999 Pluto was not the outermost planet, its eccentric orbit making Neptune the furthest from the Sun. In February 1999, Pluto again attained its status of the most distant.

That would only be a factual statement if there were no other major body yet awaiting discovery out beyond Pluto's orbit. Since 1992 astronomers have spotted some hundreds of minor planets in the region between about 30 and 60 astronomical units from the Sun, members of what is called the Edgeworth–Kuiper belt, recognizing the scientists who suggested their existence more than four decades before the first of them was discovered. Another collective name for them is the trans-Neptunian objects (TNOs). Pluto is about 1,410 miles in diameter, and is generally classed as being a major planet, the ninth in the Solar System. These numerous TNOs are mostly between 200 and 300 miles in size. In the year 2000 a new TNO was found that may be as much as 600 miles across, perhaps even larger than Ceres, the biggest minor planet (or asteroid—the terms have the same meaning) in the main belt. The nature of the TNOs seems to be quite different from the asteroids in the inner Solar System, however. Those appear to be rocky and metallic in composition, whereas TNOs are largely icy, like Pluto itself. In many ways it might be better to think of TNOs as being giant comets, thankfully keeping their distance from us, rather than classing them as minor planets.

Clyde Tombaugh discovered Pluto in 1930. For some years he had been diligently scouring photographs of the deep sky, at the Lowell Observatory in Arizona, before he eventually found a tell-tale moving point of light. At first Pluto was thought to be much larger than is actually the case, with a mass perhaps six times that of the Earth. Over the seven decades since our estimates of its dimensions have systematically downgraded it, and only recently have its mass and diameter been determined properly from eclipse observations. We start this chapter by considering Pluto's eclipses, and then see how the basic techniques employed can be extended beyond the Solar System.

THE DISCOVERY OF PLUTO

Although the perseverance with which Tombaugh searched the sky and eventually turned up Pluto is laudable, the discovery was really a fluke.

A century ago Percival Lowell and others were convinced there must be another large planet awaiting discovery, because the observed paths of Uranus and Neptune were discrepant, their positions wandering slightly away from calculations based upon the orbits and masses of the other known planets. We have seen that in the 1840s Le Verrier and Adams had successfully predicted the existence of Neptune from such meanderings of Uranus. What Lowell did was to extend this thought process, imagining evidence for some undiscovered planet.

Lowell was prone to be over-enthusiastic in his astronomical interests. In the late nineteenth century the popular idea of life on Mars was triggered to a large extent when he argued that markings on the surface of that planet were evidence of a civilization thriving there. In part his imagined "Martians" stemmed from a misinterpretation of the writings of Giovanni Schiaparelli, the Italian word for "channels" being taken by Lowell to mean "canals." River channels, of course, are natural hydrological features, whereas canals are artificial. Lowell was soon drawing Mars crisscrossed with a vast canal system. These perceived straight lines—which do not actually exist—suggested to Lowell and his followers that intelligent life existed on the red planet. They were wrong. This was a case of mass delusion.

Turning his enthusiasm to the possibility of an unknown planet beyond Neptune, the search Lowell sponsored did not bear fruit until well after his death in 1916. Even then his interpretation of the observed phenomena proved incorrect. In the decades after its discovery, astronomers realized that Pluto could not be responsible for the perceived wobbles in the orbits of Uranus and Neptune, and a resolution of that quandary did not come until the early 1990s. We will describe that solution at length, but first we must discuss how Pluto's mass was determined.

PLUTO'S MOON AND MASS

When Pluto was spotted, after such a long quest, it hit the head-lines worldwide. From its apparent brightness (or perhaps we should say its "faintness") it was obvious Pluto must be small, and some of the euphoria abated. From time to time astronomers would again turn their telescopes towards Pluto, but it was hardly in the news again until 1978 when it was found to have a moon of its own. That moon was given the name Charon (see Figure 14-1). It is about 730 miles in diameter, around half the size of Pluto itself.

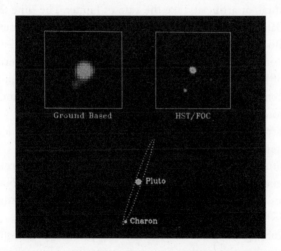

FIGURE 14-1. Images of Pluto and its moon, Charon. Ground-based observations demonstrated the latter's existence (top left), but they are only well separated by the Hubble Space Telescope (top right). Between 1985 and 1990 our edge-on alignment to their mutual orbital plane (bottom) led to repeated eclipses, allowing the sizes and combined mass of the two to be evaluated.

The finding of Charon was nice in itself, but also quite a handy thing because it allowed astronomers to determine Pluto's mass properly. Until then all that could be done was the definition of limits on its bulk from various perspectives. Taking characteristic values for the average albedo (the fraction of sunlight reflected), size limits could be calculated. Then, assuming Pluto to be made of rock, or of ice, or of a mixture, possible values for its mass could be calculated. Similarly a maximum value for its mass could be ascribed through the *lack* of major perturbations of the paths of the outer planets.

To get a better evaluation a probe is needed. That probe might be natural, or artificial. Take the case of Jupiter. The time that the four Galilean satellites take to circuit that planet can be measured, and also the sizes of their orbits. Those two pieces of information—orbit size, orbital period—make it possible to derive the mass of Jupiter, using Kepler's laws of orbital motion. Even if only one such moon existed, the Jovian mass could still be found. Having those four bright satellites makes it a cinch, because you can compare the values obtained using each of them and look for consistency in the result.

Saturn was studied in the same way, especially through its large moon Titan. When the tiny Martian moons Phobos and Deimos were identified in the late nineteenth century it became feasible to reckon the mass of the red planet.

Unfortunately, Venus and Mercury presented long-standing problems because neither has a natural satellite. Using the magnitudes of their mutual orbital perturbations, limits had been placed on their masses, in line with those expected given densities characteristic for rocky bodies with iron cores. Better evaluations awaited visits by space probe to those planets. The mass of Venus was determined in the 1960s through radio tracking of various spacecraft. Mercury had to wait until the mid-1970s, when NASA's *Mariner 10* satellite flew past it three times.

When Charon was discovered it was at last possible to have a stab at Pluto's mass, but the situation was complicated. Firstly, the two objects are close to each other, and far from the Earth. Measuring their separation was therefore extremely difficult, especially through the blurring effect of our atmosphere, although the Hubble Space Telescope improved matters (compare the two upper images in Figure 14-1). Secondly, with a mass ratio of about eight to one, the Pluto–Charon system represents a binary planet (as discussed in the Appendix). Because of these factors, interpreting the orbits to get the mass of Pluto presents difficulties.

The situation was saved by the study of their mutual eclipses. Pluto and Charon rotate every six days about their barycenter with a separation of around 12,200 miles, like a cosmic dumbbell. The scale involved is shown by the lower image in Figure 14-1. The orientation of their axis of rotation is preserved, analogous to a gigantic gyroscope, meaning that at certain times we look edge-on along the plane of their mutual orbit. This means that eclipses will occur, Charon first skimming in front of Pluto, a little over three days later passing behind it. Given the sizes of the objects, one can calculate that such sequences of eclipses will last for about five years, but in episodes separated by 124 years (half the time it takes the Pluto–Charon pair to orbit the Sun).

By a great stroke of fortune, Charon's discovery came with perfect timing, just as a five-year eclipse sequence was about to commence. This ran from 1985 to 1990, allowing astronomers to observe these events and determine a great deal about the double planet. The total intensity of light received by our telescopes was found to drop off by about 20 percent when an eclipse occurred, Charon either dipping behind Pluto or covering part of its disk. The accurate timing of the way in which the brightness varied made it possible to calculate their mutual orbits, and so their combined mass. (Note that I wrote *combined* mass there, because even these observations have ambiguities, and do not render the individual masses with precision. Charon's bulk seems to lie somewhere between 8 and 16 percent that of Pluto, but we cannot be sure.)

A better idea of the pair's characteristics is unlikely to be obtained until the first spacecraft arrives there. A probe called *Pluto Express* is on the drawing board. It would certainly need to be an express, using a gravity assist from Jupiter to speed it on its way, because a slow trajectory to distant Pluto would mean that the scientists involved in its planning would have retired before their progeny arrives at its destination. At the time of writing it seems unlikely that we will deliver any space probe to Pluto before 2016.

AN OCCULTATION BY PLUTO

The atmosphere of Pluto might also be detected using eclipses: eclipses of the third kind, starlight being used to probe its properties. If this little body has any atmosphere, then one would expect it to be most abundant when near perihelion, because the increased solar heating may cause any volatile ices to sublimate. Pluto's composition seems to comprise about 70 percent rock and 30 percent ices. The latter would be mostly water ice, but also other highly volatile solid materials like carbon monoxide, nitrogen and methane, which could form a temporary atmosphere whenever Pluto makes its nearest approach to the Sun, albeit at a distance of over 29 astronomical units.

Pluto was near perihelion in the late 1980s, and again fortune blessed astronomers interested in this little planet. In June 1988 it passed over a relatively bright star, making occultation observations feasible. The gradual dimming of the starlight before and after total obscuration allowed Pluto's tenuous atmosphere to be fathomed. There is another reason for any space mission to Pluto to be an express: as it recedes from the Sun, that atmosphere will freeze once more, leaving the planet naked for the next two centuries.

NO PLANET X

Knowing that the mass of the Pluto–Charon double planet is small—only one part in 400 of the terrestrial mass—it was clear that the apparent wanderings of Uranus and Neptune required an alternative explanation. Many have seized upon the notion that indeed the discovery of Pluto was by chance, even if part of a deliberate search, and there must be another massive body out there, a Planet X.

This idea is too simple, though. The situation here is similar to when Le Verrier tried to explain the motion of Mercury using hypothetical bodies near the Sun. No *single* unknown object could explain how Mercury moved and, while popular imagination focussed upon a planet Vulcan, Le Verrier himself knew that several intramercurial bodies would be necessary. Turning to the cases of Uranus and Neptune, again no single unobserved body could explain the apparent anomalies. There would need to be not only a Planet X, but also a Planet XI, XII, XIII, XIV, XV, and so on.

Are these indeed the minor planets now being spotted regularly out beyond Neptune? The answer is *no*—for several reasons.

One is that they are simply not big enough, being smaller even than Pluto. Another, paradoxically, is that there are too many of them. When there are many objects, all separately producing gravitational tugs—and we think that there are some millions of minor planets in that belt concentrated between 30 and 60 astronomical units—their effect is smeared out, and no distinct wobbles to the motions of the outer planets would be produced.

This was all a bit of a tease to astronomers, the solution eventually being reached only in the early 1990s. When *Voyager 2* flew by Uranus in 1986 and Neptune in 1989, radio tracking of the bending of its trajectory by the gravitational attractions of those bodies allowed researchers to determine the planetary masses with unprecedented precision. And they showed the previous values to be wrong.

When the measurement of the masses of planets using observations of natural satellites was discussed above, I did not mention Uranus and Neptune. The masses of those planets had indeed been evaluated in that way, each of them possessing a flotilla of moons, but remember that they are a long way from us. One could *time* the orbits quite accurately, by observing eclipses perhaps, but measurements of the *sizes* of their circumplanetary loops must be inherently inaccurate from this distance. However, not only did *Voyager 2* pass close by those planets, but also the radio tracking could be carried out with great precision.

When the spacecraft data were analyzed, it was realized that the previous masses for Uranus and Neptune were each slightly wrong, by a fraction of 1 percent, one too high and one too low. The improved evaluations for the masses were plugged into the numerical models for the whole Solar System, and when that was done there no disagreement remained between the observed planetary positions and the theoretical positions from the computed ephemeris. The earlier theoretical positions were wrong simply because they were based on slightly incorrect planetary masses.

By 1992 the last nail was hammered into the coffin of Planet X. The observed motions of the outer planets are consistent with there being no other planet comparable in size to the Earth anywhere within 100 astronomical units. This just shows again that the discovery of Pluto was a fluke, resulting from inaccurate data. No one was to blame for this; one must remember that there is *never* absolute certainty in science, only limits of confidence that depend on the accuracy of the information in hand at any time. Lowell and his colleagues started looking for Pluto due to wishful thinking, rather than a sober analysis of the situation, and so it was actually a happy chance that the planet was found. If Tombaugh had not spotted it in 1930, someone else would have done so before too many years were out.

ECLIPSES ELSEWHERE

It will be more than a century until we have another opportunity to witness Pluto–Charon eclipses. That pair comprises a binary planet with a mass ratio of about 8:1. Looking among the major planets, the next highest primary-to-secondary ratio is represented by the Earth and the Moon, weighing in at 81:1, so we may be justified in thinking of the Earth–Moon system as another binary planet.

Other binaries are known in the Solar System. Most asteroids are very irregular in shape (recall Figure 12-1). If they spin fast enough, asteroids may separate into component blocks that would then loop around each other, in a temporary gravitational embrace. ("Temporary" has an astronomical meaning here: it might be a million years before some close passage by a planet causes a separated asteroid to lose its grip on the fragments.) In 1993 the *Galileo* spacecraft, while on its voyage to Jupiter, visited the large asteroid Ida. It was a surprise to many that Ida was found to have a small moonlet, which has been named Dactyl. The large mass ratio, however, means that we cannot really claim Ida and Dactyl to represent a binary asteroid.

From various lines of evidence there has long been a suspicion that there are binaries among the asteroids that cross the Earth's orbit. For instance, several of the impact craters on our planet seem to be arranged in pairs formed at the same juncture. Direct evidence for such a binary object came in 1997, when observers following the brightness variation of the asteroid named Dionysus detected dips in its intensity curve characteristic of repeated mutual eclipses and occultations. Dionysus is a binary, with one lump larger than the other, and eclipses tell us so.

The astronomical context in which the term "binary" appears most often is far beyond the Solar System, in the description of binary stars. Such pairs are a well-known phenomenon. A large fraction of the apparent pinpoints of light one can see in the sky using only the naked eye actually display a dual nature if they are viewed instead through a telescope with sufficient resolution. For instance, the national flags of Australia and New Zealand both show the stars of the constellation known as the Southern Cross (or Crux), but the depictions are inaccurate on two counts. First, the colors are radically wrong. More important here, the binary properties are not shown on the flags. The brightest star in the Southern Cross is a multi-colored *triplet*, and the next brightest is a doublet. Similarly Sirius, the most luminous star in all the heavens, actually has a faint companion.

It was not until the early nineteenth century that the binary nature of many stars was widely accepted, despite earlier evidence for their existence. The largest available telescope at the time was that of William Herschel, which he said could not separate stars into discrete components. Observers reporting that some bright stars had luminous companions tended to be ridiculed. It was not until 1802 that Herschel agreed that binaries existed and could be distinguished telescopically. His son John spent much of the period between 1820 and 1840 drawing up catalogues of binary stars, initially with James South in London and Paris (recall our discussion of South in Chapter 12). John Herschel later took his family with him to Cape Town in South Africa, from where he scanned the southern sky for binary stars for four years.

These were *visual binaries*—stars that could be resolved by eye using a good instrument. Nowadays astronomers study more distant binary systems, too far to be separated directly, by analyzing their composite spectra. The spectrum emitted by each of the two stars, often of quite disparate types, will display varying red- and blue-shifts as first one star and then the other approaches and recedes from us in their locked orbits about the mutual center of gravity. (The speed-dependent shift in the spectrum of an object, due to the Doppler effect in light, was discussed in Chapter 12.) Such stars are called *spectroscopic binaries*. As with Pluto and Charon, those orbits allow the stars' masses to be investigated.

The first binary star to have its physical properties probed through such orbital data was not a spectroscopic binary, though. The star in question displayed not spectral changes, but rhythmic variations in its intensity.

THE ECLIPSES OF ALGOL

If you overshoot the mark when looking up the word "algol" in a dictionary, you may be surprised. Algology is the study of algae. Algolagnia is a psychiatric term covering sadism and masochism. ALGOL is an acronym recognized by the computer-literate, standing for ALGOrithmic Language, one of the earliest programming codes. It is Algol, a capitalized proper name, which is the subject of our present inquiry. This is otherwise known as *Beta Persei*, the second-brightest star in the constellation Perseus. ("Algol" is an old Arabic word, apparently meaning "demon," so perhaps the ancients recognized its peculiar behavior long before tardy Western science did so.)

This star's significance stems from being the first to be identified as an eclipsing binary, although its true nature was not widely comprehended until two centuries after 1667, when its radically-varying brightness first had been noted in the post-Renaissance era. In that year Geminiano Montanari, who was examining the sky from Bologna in Italy, recorded that at times it appeared much fainter than normal. No telescope is required to see this, just good visual acuity and patience.

After that no further notice was paid to Algol until 1782 when John Goodricke systematically followed its brightness over an extended period. Goodricke was an English astronomy enthusiast, a deaf-mute and just 18 years old at that stage. He found that the star's apparent brightness decreased over several hours and then enhanced again, this trend being repeated every 69 hours, as regularly as clockwork. Goodricke communicated his discovery to the Royal Society of London and hazarded the guess that the variability might be due to some unseen pale object orbiting the star—a planet perhaps—or possibly spots like those on the Sun, quickly moving across its surface. The notion of binary stars was yet unsuspected.

A few years later a Swiss mathematician, Daniel Huber, used Goodricke's observations to show that "star spots" could not be responsible. He suggested instead that Algol had a darker companion star and, from an analysis of the way in which the brightness varied in time, was able to derive both feasible sizes for the pair and their separation from each other. This was pioneering work in what later became a standard field of astronomy.

Others were also of the opinion that Algol must be a binary star system producing regular eclipses but, strangely, even after the existence of visual binaries was accepted in the early nineteenth century, still the case of Algol lay dormant. Variable stars were seen, but not with the same form of brightness fluctuations as Algol. In science it is frequently the case that discrepant observations are ignored, because they do not fit in with mainstream thought at the time, and may lay ignored for years, or even decades. This was certainly the case with Algol. It was not until much later, when several similar cases were recognized, that the concept of eclipsing binaries gained a foothold, a hundred years after Goodricke and Huber got an inkling of the explanation. For them the trail did not go further, and it was long after their deaths that other astronomers realized they had been correct.

Eclipses by the Algol binary system are interesting because they differ in a fundamental way from all the types of eclipse previously mentioned. Algol comprises two large stars, one about three times the solar diameter, the other four times. They produce eclipses, as shown in Figure 14-2, which superficially might be considered similar to those of Pluto and Charon, but actually they

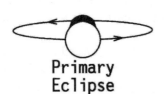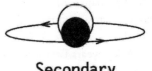

Primary Eclipse **Secondary Eclipse**

FIGURE 14-2. The eclipses of the Algol binary star system. Because the smaller star is much brighter than the larger, the primary eclipses cause the overall intensity to dip by a factor of three, while the secondary eclipses result in dimming by only about 10 percent.

are very different. Planets and their moons cannot *produce* light. All that Pluto and Charon (and all the other objects in the Solar System) do is *reflect* the light of the Sun back from their surfaces. During a Pluto–Charon eclipse the drop in the total intensity we receive is only 20 percent, depending solely on their comparative areas and albedos. In contrast, in a binary star system both components *emit* their own light, making possible much larger amplitudes in the variation of the total light received in our telescopes.

In most binary stars, the members are of differing types, and the intensities of the light each emits depend on the complexities of stellar evolution and their internal workings. It does not follow, therefore, that the larger of two stars must be the brightest, or even have the greater mass. If anything, the converse tends to be the case. A more massive star will have greater self-gravity, which condenses it. This makes it hotter and denser in its interior, promoting nuclear fusion (the energy generation within stars through fusion reactions was discussed in Chapter 5). As a result, its energy generation rate would be elevated. Smaller stars tend to be hot and thus white in color, larger ones cooler and redder. The amount of emitted light rises as the fourth power of the surface temperature, overpowering the influence of the greater surface area of big stars. It is like comparing a red-hot poker pulled from a fire with the filament in an electric light bulb: there is no doubt concerning which is the brighter. The tiny filament emits more light because it is much, much hotter.

In the case of Algol, then, the smaller member is hotter and brighter than its relatively dim companion. In the secondary eclipses (see Figure 14-2), at the phase when about half of the larger member is covered by the more brilliant, the total brightness of the system falls by only about 10 percent. In contrast, in the primary eclipses the dim star obscures more than half of its brighter companion, and the total intensity plummets by a factor of three. Separated by 69 hours, such eclipses last for 10 hours from start to end, the faintest part persisting for only an hour or so. During a long winter night Goodricke or others might have witnessed a complete eclipse in the Algol system and easily charted its relative magnitudes by comparison with other stars. Three nights later they could have seen the same thing, although gradually the eclipses would have fallen back until they occurred in daytime, because the cycle is not a multiple of 24 hours. Timing of the eclipses over a month or so, when visible, would have allowed the astronomers to determine the consistent 69-hour period governing the eclipses.

The fact that Algol's errant behavior is so obvious, and yet was ignored by the astronomical establishment for many decades, is a prime example of scientific conservatism. Scientists are some of the most conventional of creatures, the majority being totally unwilling to stick their necks out. Thinking back to the lead-up to the outermost planet's discovery, one might poke fun at Percival

Lowell and his beliefs about life on Mars, and the basis of the search that fortuitously turned up Pluto. Then again, I reckon that he derived more enjoyment from his astronomy than those who criticized him, before his death and after. "It takes all sorts to make the world turn," goes the old aphorism, and the thought may be extended to the entire universe, and our study of it. Without the radicals who will not listen to "conventional wisdom," scientific progress would be even slower than it is now.

15

An Eclipse Chaser's Guide

High on her speculative tower
Stood science waiting for the hour
When Sol was destined to endure
That darkening of his radiant face
Which Superstition chose to chase,
Erstwhile, with rites impure.

William Wordsworth, *The Eclipse of the Sun*

H aving reached this point in the book, it almost seems superfluous to mention what happens during a total solar eclipse. Indeed this was not intended as an eclipse watcher's handbook, but rather an extended account of just why eclipses have been important in the development of human civilization. Nevertheless we should mention some of the phenomena: for completeness and interest. We must start with the safety aspects.

SAFE SOLAR ECLIPSE VIEWING

Typically the first contact, when the Moon begins cutting a notch from the solar disk, occurs about 75 minutes before totality, giving you an extended period during which the movement of the Moon across the face of the Sun may be monitored. How can you view this? Not with the naked eye, and certainly not through any optical device like binoculars or a telescope. At the time of the eclipse in August 1999 over England and much of Europe, the newspapers were full of warnings, telling people of the danger posed to their eyesight. This led one letter writer to the London *Times* to suggest that "In view of the many warnings regarding adequate safety measures when viewing the eclipse, should we not be sensible and listen to it on the radio?" The following malapropism is especially delightful in that it appeared in the *Lady*, a hugely staid and strait-laced British magazine: "But few seem to realize that looking at an eclipse with the naked is dangerous . . ." The danger attached to eclipse watching though is a serious matter.

Almost a millennium ago, Al-Biruni, a multitalented Islamic scholar from the lands south of the Aral Sea, warned: "The faculty of sight cannot resist it [looking at the Sun directly], which can inflict a painful injury. If one continues to look at it, one's sight becomes dazzled and dimmed, so it is preferable to look at its image in water and avoid a direct look at it, because the intensity of its rays is thereby reduced. Indeed such observations of solar eclipses in my youth have weakened my eyesight." In the eleventh century Al-Biruni did not have the advantage of either a telescope to project an image safely onto a screen or optical filters. His suggestion was to view the Sun reflected from the surface of water in a bowl, which (depending upon the angles involved) can result in a few percent or less of the sunlight reaching the eye. This was a trick employed far back in antiquity by the Babylonians, Egyptians, Romans, and Greeks alike, the smarter ones using oil or pitch because their high viscosity makes for less rippling. There is no need to resort to such outmoded techniques nowadays; simply use a filter.

What sort of filter is needed? One simple filter is a piece of grossly overexposed black-and-white film that has been fully processed. This leaves it largely opaque and if you peer through this then perhaps just one part in 10,000 of the sunlight makes it to your eye. Note that only certain types of black-and-white film will do: it is the silver granules that block most of the sunlight and the dyes used in color film are *not* adequate. There are also potential drawbacks with this method, such as the possibility of scratches through the emulsion allowing too much light to strike your eye. Similarly, smoked glass is inadequate and dangerous.

These, then, are cheap but unsatisfactory solutions. Bear in mind the various aphorisms along the lines of "don't spoil the broth for want of a pinch of salt," as this is a case where economy may lead not only to the broth missing its salt, but being poisoned with arsenic to boot. Don't take silly risks for the want of a proper filter. There are many available commercially at little cost, often in the form of goggles with paper frames and flat "lenses." These seem totally black until you look through them at a very bright source, and find that just a tiny fraction of the light penetrates the filter: just enough to enable you to monitor the progress of an eclipse safely.

Amateur astronomers usually have large filters to fix over the openings of their telescopes, allowing direct viewing, but unless you know precisely what you are doing, never put your eye near the ocular of any instrument directed towards the Sun. A projected image may be obtained using a small telescope (as in Figure 1-13), or a pair of binoculars clamped in a stand and with one lens covered, just in case. Often the image is so bright on its screen that it is necessary to stop down the aperture, by covering the top end of the telescope with a card penetrated by a suitably small hole, allowing only a fraction of the impinging sunlight to enter the instrument.

The partial phase of the eclipse can be followed using some sort of pinhole camera, such as a shoebox with one end cut out and a partially translucent paper screen taped in its place and a small hole punched in the opposite end. It would be even simpler to use a small mirror as follows: cut a hole about a quarter-inch across in a sheet of card, fix that over the mirror's surface, and reflect the sunlight coming through the peephole back onto a shadowed wall. This will form an image of the solar disk, the lunar notch enlarging and creeping across it. Breaking a mirror is considered unlucky by the superstitious, but deliberately smashing one may be a good idea for an eclipse because each fragment may be used to reflect the sunlight and produce an image of the partial phase.

Actually, no equipment at all is needed to observe the partial eclipse. I often tell people to think of the surefire cure for seasickness, and also to look at the ground, not the sky. What is the cure for seasickness? Sit under a tree—it always works. If you are positioned under a suitable tree, with dense foliage, and look at the ground, you will see that the tiny gaps between the leaves act as natural pinhole cameras, casting myriad crescent images all around you. An example is shown in Figure 15-1.

To look directly at the Sun during the partial eclipse, on go your eclipse-viewing filters. The only time it is safe to view the Sun without such equipment is during totality, when your goggles or whatever equipment you have been using should be removed, else you will miss seeing the best bits. Apart from the short phase of totality—that precious couple of minutes—you must always have an appropriate filter to protect your eyes, if you want to gaze directly at the Sun.

FIGURE 15-1. The foliage of a tree provides a set of natural pinhole cameras, producing crescent images during the partial phase of a solar eclipse.

THE ECLIPSE PROGRESSES

As the partial phase progresses, you are moving deeper and deeper into the Moon's penumbra, as sketched in Figure 2-3. In Figure 2-4 we saw the lunar shadow cast on a largely cloud-covered globe in August 1999, as photographed from a low orbit above the atmosphere. Better images, of the annular eclipse in February 1999, are shown in Figures 15-2 and 15-3. These show the shadow over Western Australia, the coastline of that country plus parts of Southeast Asia being obvious.

In the last 10 to 20 minutes prior to totality the ambient light diminishes considerably. Not only its intensity alters, but also its tone, obtaining an eerie quality and a grayish hue, almost metallic in guise. As Percy Bysshe Shelley wrote:

FIGURE 15-2. An annular eclipse swept across Australia on February 16, 1999. This image, obtained by the Japanese high-orbiting GMS-5 satellite, shows the globe soon after the shadow entered Western Australia, leaving that area much darker than the similarly cloud-free regions of Southeast Asia visible further to the north.

> With hue like that which some great painter dips
> His pencil in the gloom of earthquake and eclipse.

Some people report that a green coloration appears, but that is generally because they have looked too closely at the Sun itself (recall the quote from Shakespeare's *Taming of the Shrew* in Chapter 12). Way back in 1185 an eclipse viewed in Russia produced this report: "On the first day of the month of May, during the ringing of the bells for the evening service, there was a sign in the Sun. It became very dark for an hour or longer and the stars were visible

FIGURE 15-3. This image obtained with the _NOAA-14_ meteorological satellite shows the lunar shadow over Western Australia in more detail. Although there were banks of cloud to the far north and south, the many observers concentrated just below the town of Geraldton, where the eclipse path met the coast, had clear skies. This picture was obtained a few minutes later, when the whole shadow was over land.

and to men everything seemed as if it were green. The Sun became like a crescent of the new moon and from its horns a glow like a roasting fire was coming forth." One must avoid affecting one's eyes in this way because it takes some minutes for them to recover and by then the totality will be over. Appropriate goggles will do the trick.

It is at this stage of gathering darkness that animals (and some humans) start to get confused. Birds land in the trees and go quiet, their anxiety being palpable. Conversely insects start to scrape and sing, as they do at dusk. Bats and nocturnal moths take to the wing, while butterflies settle and flowers begin to close their petals. Dogs may start to howl. Bees can get especially confused because they navigate by the polarization of the sky, and that depends on the angle of the Sun. Similarly, people may be psychologically affected in various ways, few being left unmoved by the experience of totality. That is very much an individual thing.

TOTALITY APPROACHES

As the obscuration of the Sun increases the sky darkens, although it never gets as black as dead of night. That would be too humdrum. The sky qualities during an eclipse are much more intriguing and unusual than this.

First we should think about what can be seen *because* the sky is dark. Many people seem to believe that no stars exist during the day, but they are there, simply drowned by the bright blue sky. If you don't believe me, arrange to use a telescope one clear day and be sure to avoid pointing it at the Sun. The stars are there and of course with the naked eye the Moon is also often visible. Similarly, if you know where to look then Venus can be viewed unaided during daytime, although because of its orbit it's always quite near the Sun, which is why one sees it best either soon after sunset or just before sunrise.

Similarly Mercury always stays close to the Sun, and many people only consciously spot that planet *during* an eclipse. I write "consciously" because it is often seen and yet not recognized by the viewers. Many have sat and watched the Sun go down in the west over a placid ocean, and then wondered about a bright, slightly reddish "star" just above the horizon. If you've done that, chances are you've seen Mercury. My favorite memory of the type is having sat in a Jacuzzi at a splendid house on Malibu Beach with a movie producer friend, and after the sky had darkened still more we could see Comet Hale–Bopp blazing across the firmament.

The other planets though also move across the sky on paths close to the ecliptic. Depending upon the particular eclipse, it's likely that you'll have Mars, Jupiter, or Saturn providing a celestial jewel or two to glitter and attract your attention. The bright stars will also be out to dazzle you as the sky darkens, the specific array depending upon the season. Maybe it will be Castor and Pollux, the Gemini twins, accompanied by such stellar beasts as Sirius, Procyon, and Capella. But all of these are available at some time of the year during clear nights. If you're blessed with cloud-free skies for an eclipse, it is the special phenomena that should occupy your attention.

Let us imagine that totality is now imminent, a few minutes to go. The temperature is dropping perceptibly, and many watchers start to shiver (so take a sweater). An effect often glimpsed just fleetingly is the *shadow-band phenomenon*. Turbulence in the Earth's atmosphere causes differential refractive effects (bending of the paths taken by light), which is why the stars twinkle, as discussed in Chapter 12. The planets, however, look bigger because they are much closer to us and so do *not* twinkle. This is an easy way to differentiate Mars or Saturn from the stars at night. The Sun is normally much too large to twinkle, but as totality approaches only a slender crescent of the solar disk is left, making the equivalent of twinkling possible, except that here we have a very bright

FIGURE 15-4. The shadow band phenomenon sketched, with some imagination, after an eclipse in Spain about a century ago. °

source. If the conditions are right then you may see wavy bands of light flickering quickly over the terrain; their viewing is easier if you have something like a large white sheet spread over the ground. These shadow bands are similar to the patterns seen on the bottom of a swimming pool, except with much less contrast (they vary in intensity by a few percent at most). Photographs of these bands have proven elusive, with few clear examples. A sketch, drawn with very considerable artistic license, is shown in Figure 15-4.

The Moon's shadow traverses the Earth at about 1,600 miles an hour. During the partial stage the increasing penumbral penetration is not noticeable on a minute-to-minute basis, but as the umbra approaches things start to happen fast. The complete lunar shadow can be seen zooming towards you from the west like a vast storm bearing down at supersonic speed. An elevated viewing location with a clear horizon to the west has much to recommend it, such that the rapidly encroaching shadow may be seen in these last 10 to 20 seconds before totality.

There are other aspects of the shadow to note. Totality only takes place within the narrow band that you have sought out, and a few tens of miles to the north or the south there is incomplete blanking of the Sun. You can see the sky that far away—looking beyond the edge of the shadow—and it will appear the same orange as twilight, eventually all around the horizon.

Now to the Sun and Moon themselves. In the last quarter-minute Baily's beads appear around the lunar limb, the final few specks of light passing between the mountains of the Moon, these seeming to shift around the periphery of the disk until only one is left: the diamond ring effect. A few more seconds and it is gone. That's second contact. Totality is with you.

THE PHASE OF TOTALITY

As totality begins, the first thing to note is the chromosphere, as discussed in Chapter 5. The chromosphere is seen as a pinkish region (hence its name) along the limb near where the diamond ring just blinked out. It comprises a layer about 2,500 miles thick above the photosphere, but so much less intense that it cannot be seen except during an eclipse.

The corona, a pearly white crown extending several solar diameters above the surface, may have been apparent in the minute before second contact. Typically the corona is a million times fainter than the solar surface, which is why it cannot be seen except when the photosphere is mostly extinguished. The form of the corona varies with the solar cycle, which had a peak in 2000/2001. When the Sun is very active, a complete white aureole may occur, rather than the patchy corona with significant concentrations—the plumes and streamers—seen during periods of lower activity (as was portrayed in Figure 1-3).

Prominences may or may not be present. Figure 1-5 shows rather vividly that such structures are transient, often lasting only hours or days. Like the weather on any date, they cannot be predicted until, at best, the day before an eclipse. If there are any present, then the nineteenth-century term for these loops and arcs—the *red flames*—provides a pretty good summary of their appearance. Prominences may snake above the surface by a third or more of the solar radius.

Some solar eclipses produce totality for as much as seven minutes (such as those indicated in Figure 2-2), but typically the period is between two and three minutes. Some people experience that as lasting an age; for others it is come and gone in no time at all. Charles Lambert, a member of the French eclipse expedition to Sudan in 1860, had this to say: "But at the moment of totality, all became silent and dumb. Neither a cry nor a rustling, nor even a whisper was heard, but everywhere there was anxiety and consternation. To everyone the two minutes of the eclipse were like two hours." On the other hand British astronomer Edward Dunkin, who went to northern Scandinavia to observe an eclipse in 1851, was frustrated by the brevity of totality. "So absorbed was I during this short interval that when the limb of the Sun reappeared I could scarcely realize the fact that two and a half minutes had elapsed since the commencement of totality. These were truly exciting moments, and although I had hastily witnessed most of the phenomena, I felt somewhat disappointed that more had not been accomplished. Few can imagine how much I longed for another minute, for what I had witnessed seemed very much like a dream." Things are hectic during the hundred seconds or so of total eclipse with which one may be blessed. Keen amateur astronomers tend to record dictation tapes ahead of time, with countdowns for what they need to do to get all their planned photographs. Activity is frenetic and it's easy to get caught up with just staring, missing some of the things one might like to note while there is the fleeting opportunity.

Totality ends with third contact, when the diamond ring appears again. For the few minutes of totality, the eclipse should be viewed without filtration, but those goggles need to be on again for when the solar surface flashes back into view. Apart from perhaps damaging your eyes, the unattenuated brightness striking your retina will limit your ability to see Baily's beads clearly. You might also miss the subsequent phenomena, such as the lunar shadow rushing eastwards as the Moon withdraws from the Sun. Then there is another hour or so of partial eclipse until fourth contact, when the Moon ceases all overlap with the solar disk, but that of course is all rather anticlimactic.

FROM TIMES PAST TO TIMES FUTURE

We have described above what can be seen during a total solar eclipse and various past eclipses were mentioned in passing. With luck, this will have whetted your appetite and you'll be hungry to experience one yourself. So let us see what the future has in store for us.

Ancient sky watchers were able to predict the future—to some extent—using the tapestry of eclipses described in Chapter 3. There we presented two sketches covering all eclipses between 1900 and 2100: solar events in Figure 3-1, and lunar in Figure 3-2. Both types follow the same basic rules and so produce similar patterns. The short-term sequences of solar eclipses have greater

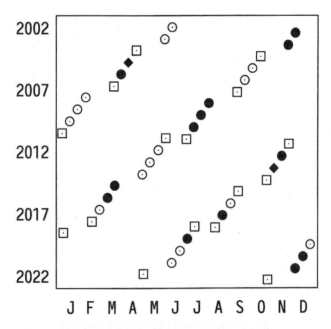

FIGURE 15-5. Plotted here are the dates for all solar eclipses due between 2002 and 2022. Solid circles represent total eclipses, open circles annular eclipses, and black diamonds hybrid events (part annular/part total). Partial eclipses are shown as squares.

numbers of members than the lunar because the ecliptic limits (see the Appendix) are more stringent for the latter.

Now we require a more detailed view, to show the eclipses due over the next two decades. By extracting the pertinent data and plotting them again, in a slightly different way, Figures 15-5 and 15-6 result. Those plots in hand, let us see what the heavens have in store for the eclipse watcher.

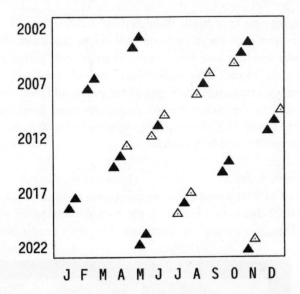

FIGURE 15-6. Lunar eclipses due between 2002 and 2022. Solid triangles represent total eclipses, open symbols are partial events. Only umbral eclipses are charted; over this period 19 penumbral eclipses will occur, but they are of little interest.

SOLAR ECLIPSES 2002–2022

From 2002 to 2022, 46 solar eclipses will occur: 13 total, 15 annular, 2 hybrid (changing between annular and total along the track), and 16 partial. The total eclipses are the gems, and the major quest of enthusiasts, and so we concentrate upon the 15 events producing at least some period of totality.

The years 2002 and 2003 each will have one total and one annular eclipse, before 2004 has partial eclipses only. This basic form happens to continue through these decades: a run of two or three years each containing one total eclipse and most often an annular one, too, and then a year containing only partial events. There are 15 upcoming opportunities to see a total eclipse, and below we summarize when and where you should place yourself in order to experience the stunning phenomena firsthand. The map in Figure 15-7, showing the ground tracks for all such eclipses between 1996 and 2020, will help.

December 4, 2002:
In June of 2001 a total eclipse visited Angola and again in December 2002 that nation is crossed, the track continuing on a more southerly route along the Botswana–Zimbabwe border. Most of

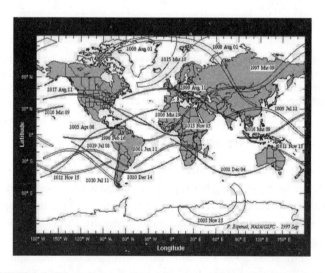

FIGURE 15-7. This world map shows the ground tracks for all total solar eclipses between 1996 and 2020.

the path is over the deep southern reaches of the Indian Ocean, but it enters land again near Ceduna in South Australia before frittering out in the northeast of that state. The maximum duration of totality is a few seconds over two minutes.

November 23, 2003:
Abandon all hope ye who enter here: the path of totality cuts only across a portion of Antarctica. Although this is near the start of the austral summer, so early in the season the sea-ice has yet to disperse sufficiently to make feasible a visit by a cruise ship to the great southern continent itself. If you can get there, totality will last for a few seconds less than two minutes.

April 8, 2005:
This is one of the hybrid annular/total eclipses, made possible by the finite size of the Earth: to begin with it is annular because the locations crossed are further from the Moon than those close to the sub-solar/lunar point near the middle of the track length. Unfortunately the portion giving totality is in the Pacific, just south of the equator, making a seaborne expedition necessary. The track there is only 15 miles wide and totality will last but 42 seconds.

March 29, 2006:
The track touches down in northeastern Brazil, crosses the equatorial Atlantic, and then enters Africa over Ghana and Togo. Continuing northeastwards it transits the Sahara before leaving the continent at the junction between Libya and Egypt. Sweeping over Turkey it traverses the north of the Caspian Sea and central Asia before terminating in Siberia just north of Mongolia. This is a

fairly long eclipse, with a maximum duration of more than four minutes.

August 1, 2008:
Starting in the far north of Canada, the track of totality crosses Greenland before descending over a Russian island called Novaja Zemla and then southeast through central Siberia. The west of Mongolia is touched before the track enters China where it terminates just before sunset. A view from the Great Wall would be splendid. The duration is almost two and a half minutes.

July 22, 2009:
This eclipse is significant as the next in the saronic sequence of long eclipses shown in Figure 2.2. The last one, in 1991, crossed Hawaii and then passed down through Central America, eventually petering out in Brazil. In 2009, the duration will be as much as 6 minutes and 38 seconds. The track begins off the western coast of India, cutting across that country before traversing the eastern Himalayas and then China again. At Shanghai it leaves land, moving out over the Pacific.

July 11, 2010:
Apart from Easter Island—another splendid place from which to witness an eclipse—this is another inhospitable event, reaching the south of Chile and Argentina close to sunset in the depth of the austral winter.

November 13, 2012:
The four-minute totality begins near Darwin in Australia's Northern Territory, then crosses the north of Queensland and the Great

Barrier Reef before heading out over the Pacific. There seems little doubt about the best place for viewing, hopefully before the rainy season starts in tropical Australia.

November 3, 2013:
The track of this hybrid annular/total eclipse begins in the Atlantic somewhat east of Florida and travels southeast, then across central Africa. There is a better opportunity to witness totality than in the case of the 2005 hybrid in that the track is wider (almost 40 miles) and totality longer (100 seconds), but again that portion occurs over water in the equatorial Atlantic.

March 20, 2015:
The track runs northeast between Scotland and Iceland, making the Faeroe Islands the only accessible land at this time of year, unless one wants to winter over in Norway's far-north Svalbard archipelago. The eclipse actually ends with sunset at the North Pole, on the first day of sunlight after the six-month winter darkness. The maximum duration is 2 minutes and 46 seconds. Its saros pair in Figure 15-7 is the eclipse of March 9, 1997.

March 9, 2016:
Sumatra and southern Borneo are the larger landmasses under the track of this four-minute eclipse, which is mostly over water. Referring to Figure 15-7 one sees how this eclipse echoes that of February 26, 1998, which was a saros earlier.

August 21, 2017:
By the time this 70-mile wide track arrives, the United States will have been waiting 38 years for a total eclipse. It hits land at Salem,

Oregon, during the morning of Monday August 21 and eventually departs into the Atlantic in the early afternoon at Charleston, South Carolina. There are many other cities along the way. The maximum duration of 2 minutes and 40 seconds will be achieved in western Kentucky. This eclipse will also be notable because the bright star Regulus will be only one degree from the eclipsed Sun, so that it will peer through the periphery of the solar corona. Mars and Mercury will also be nearby. As mentioned earlier in the book, this eclipse is the one that follows in the same saronic cycle as that of August 11, 1999, and so their tracks echo each other in Figure 15-7.

July 2, 2019:
Mostly over the southeastern Pacific, the track crosses Chile and Argentina. Except for the skiing, the high Andes is not a place to be in July, and so the coast of Chile or the pampas provide the best bets. The maximum eclipse of four and a half minutes occurs out over the ocean. Compare its track to that of June 21, 2001, in Figure 15-7 (another saros pair).

December 14, 2020:
Chile and Argentina get another chance, this time at a more clement time of year, with an eclipse lasting a little over two minutes. Again the linkage between this eclipse and its predecessor in a saronic cycle, the event of December 4, 2002, is clear in Figure 15-7.

December 4, 2021:
Applying what we know about geographical shifts after one saros and the several examples mentioned above, it is obvious that this

eclipse, linked to that on November 23, 2003, must also be over the Antarctic and so is not an attractive proposition.

So you want to see a total solar eclipse. Where should you go? Given an unlimited budget, you have a choice of locations in southern Africa or Australia late in 2002. Turkey is likely the best bet in 2006. The Great Wall of China in 2008 is a must-do, and you could go back to the coast of that country, or India, in 2009. Easter Island with its monolithic carved heads gazing perennially at the rising Sun is the only place to be for the 2010 eclipse; similarly Australia's Great Barrier Reef in 2012. For a radical climate change head for the Faeroe Islands near the spring equinox in 2015, and then don your tropical vestments again for Indonesia in 2016. After that it's Chile or Argentina in both 2019 and 2020.

An unlimited travel budget would not only be nice, but virtually a necessity if you wanted to complete that itinerary. American readers, however, have a stay-at-home opportunity in 2017, and doubtless many aficionados will have their own favored spots in mind already. My pick of a place from which to view it would the Grand Tetons. A map showing the path of that eclipse (and all other total solar eclipses crossing North America until 2050) is shown in Figure 15-8, for your own long-term planning.

The good news for the United States is that after the 38-year hiatus since 1979, when only parts of Oregon, Washington, Idaho, Montana, and North Dakota were crossed, there will only be another 7 years to wait until the next one. On April 8, 2024 there will be another total solar eclipse and it is a long one. With a 120-mile-wide track this four-minute event will pass centrally over Mexico, then Texas (including Dallas) and a chunk of the Midwest before reaching Cleveland, and then Buffalo and Montreal.

There is another peculiarity one might note about the April 2024 eclipse. Back in Chapter 11 we discussed how certain locations get more than their fair share of eclipses, focusing on Nantucket Island. We saw above that the maximum duration of the August 2017 eclipse will occur over Kentucky. Now refer to our North American eclipse map, Figure 15-8. It happens that the track for the 2024 eclipse crosses that for 2017 just to the west of there, mostly over southern Illinois, around the confluence of the Ohio and the Mississippi Rivers. This means that the good people of Carbondale will be so fortunate as to get not only the near-

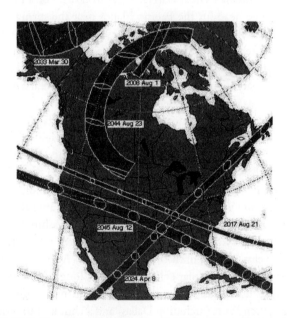

FIGURE 15-8. The ground tracks for all total solar eclipses crossing North America through to 2050.

longest eclipse totality in 2017, but also another event less than seven years later. Perhaps this is more than fair recompense for having the name of their state spelled incorrectly in Figure 10-1.

After April 2024, Alaska is the place to head in 2033 if you can stand the weather at the end of March. In August 2044 Montana and North Dakota are again lucky, although further north into Canada one will have a better view, lasting just over two minutes; this time it will be the mosquitoes rather than the snow you would need to battle. Just 354 days later, on August 12, 2045, the United States will be treated to a real humdinger of an eclipse, the track arriving over northern California and then following a track parallel to that of 2017 before it blankets most of Florida on departure. This one will be a beauty, with a track up to 160 miles wide and duration just over six minutes.

WHAT IF IT RAINS?

What if, despite your best efforts to pick the best location from which to observe an eclipse, it rains? Well, look on the bright side. A gentle sprinkle of rain will not stop *all* the sunlight getting to you. The clouds will just impede your direct view and the light of corona, chromosphere, and prominences may trickle through.

But what sort of light is that? Well, it's pink. That is, if the pluvial conditions are such that a rainbow is produced, then that rainbow will look very different from the norm. A red arc will dominate that rainbow, with little intensity in the other parts of the spectrum. Such a thing has been seen in recent decades, during an eclipse in Colombia. If all else fails one could console oneself with the knowledge that few people have ever witnessed a pink rainbow. Make sure you get a photograph.

FORTHCOMING ANNULAR ECLIPSES

Although they are by no means as striking as total eclipses, annular eclipses can afford a semblance of the experience. On June 20, 2002, the narrow path of a short-lived annular eclipse will snake its way across the northern Pacific. A better opportunity presents itself in the following year, on May 31, 2003, when Iceland, Greenland, or the Highlands of Scotland should be your destination.

On that date a most peculiar annular eclipse will occur, making it interesting in its own right. Solar eclipses can only be seen during the daytime, of course, but this one involves the Sun effectively peeking over the top of the planet. The date is just three weeks before the summer solstice, so the Northern Hemisphere is tilted almost as far toward the Sun as it goes, with the result that the Land of the Midnight Sun is indeed getting 24 hours of sunlight. *Any* solar eclipse will be visible at that time of year if you are far enough north.

In this case a partial eclipse occurs over a vast area covering Alaska, all the Arctic, Europe, Egypt, Saudi Arabia, Russia, central Asia, and Siberia, but the annular eclipse is detectable only from a restricted D-shaped region centered near Iceland. This covers some of Greenland to the northwest and to the southeast the Faeroes, Shetlands, and parts of northern Scotland.

Undoubtedly many enthusiasts will be heading for the north Atlantic region to see this event, but if you go be sure to take your alarm clock. The eclipse happens at around four in the morning, with the Sun barely above the horizon.

An even better opportunity occurs on October 3, 2005. The track then will sweep over the Iberian Peninsula and then diagonally down through Africa.

Readers in the United States who are eagerly anticipating the total solar eclipse in August 2017 might care to note that they will have a chance to practice, using an annular eclipse, in 2012. On May 20 the event will begin over southern China and then arc up over Japan and the northern Pacific Ocean before meeting the coast of North America over northern California and the south-western tip of Oregon. The path of the annular eclipse will then paint a stripe through central Nevada and other western states before petering out in northwestern Texas. The choice place from which to watch? I'd go for the Four Corners region, where Utah, Arizona, New Mexico, and Colorado meet.

Although there are more annular than total eclipses scheduled soon, it happens that overall the total eclipses are better positioned for viewing: by chance most of the annular eclipses due over the next decade are mainly over the oceans. Give thanks for small mercies.

LUNAR ECLIPSES 2002–2022

Lunar eclipses are an entirely different prospect: a good fraction can be seen without leaving home sweet home. Maps like that in Figure 2-5 are readily available, indicating that well over half of the globe gets to see at least part of each lunar eclipse. In the case of that specific eclipse—the one due on May 16, 2003—the global map shows that it can be viewed from the east coast of North America in its entirety, with locations further west than Chicago and Dallas seeing the Moon in eclipse as it rises.

In fact, reckoning whether you will be able to see a lunar eclipse is quite straightforward so long as you know when it will take place, in Universal Time (the correct term for what is often called Greenwich Mean Time or GMT). The sums are not diffi-cult. Take the example of the above eclipse. Greatest eclipse is at 03:40 UT, which tells you the central longitude of the area on the Earth from which it may be seen: 3 hours and 40 minutes, which is equivalent to 55 degrees, to the west of the Greenwich merid-ian. This longitude passes down through Newfoundland and then through the western Atlantic, eventually meeting the land again at the northern coast of South America and then proceeding south through the middle of Brazil. Anywhere within about 90 degrees of longitude of that meridian will be able to see the complete eclipse.

There is also a latitude effect, however, due to the tilt of our spin axis. This eclipse will take place five weeks before the summer solstice, when the Northern Hemisphere is tipped *towards* the Sun during the day, which means that it is tipped *away from* the Moon in opposition at night. Thus more southern latitudes are favored (as is clear from Figure 2-5), the converse being true for an eclipse during the winter. Simply put, you are more likely to see a lunar eclipse during a long winter night than a short summer night. There's not much more to it.

Unlike totality in a solar eclipse, which is brief and striking, a total lunar eclipse is more protracted, typically lasting from 60 to 80 minutes. Such eclipses are certainly dramatic in their own way, but they have neither the rarity value, nor the effects on animals and humans alike, that distinguish solar eclipses. Nevertheless they are well worth watching, when the chance arises, so let us summa-rize the circumstances for the first seven total lunar eclipses in Figure 15-6, through to the year 2010. All times given are in Uni-versal Time. In the United States one needs to knock five hours off for Eastern Standard Time, and so on through to eight hours for Pacific Standard Time.

May 16, 2003:
This is the eclipse in Figure 2-5, about which much has been said already. The map gives locations from where it can be seen: greatest eclipse at 03:40 points to the western Atlantic. It is visible throughout the Americas and in part from Africa and Europe. Totality lasts for 53 minutes. (Note also that there is a transit of Mercury nine days earlier, on May 7. This may also be seen from the eastern parts of North America, although Europe and Asia are better located because the transit straddles 08:00 UT.)

November 9, 2003:
Time 01:20 puts it over the eastern Atlantic, but totality is relatively brief (only 23 minutes). The eclipse is visible from Europe, Africa, western parts of Asia, and throughout the Americas as the Moon rises in the evening.

May 4, 2004:
This event is centered on 20:30, and so over the Middle East. It may be seen at least in part from Europe and most of Asia, plus all of Africa. It is not visible in any phase from North America.

October 28, 2004:
Another western Atlantic meridian, at just after 03:00, points to the Americas plus most of Europe and Africa having a view. Only the states on the West Coast of the United States miss any of the eclipse, the penumbral phase being in process as the Moon rises.

2005 and 2006:
As shown in Figure 15-6, there are no total lunar eclipses in either of these years, just a single partial eclipse in each.

March 3, 2007:
Occurrence at 23:20, and so ideally located for European viewers. Because this is near the equinox, the Earth's tilt has little effect, and so all locations within about 90 degrees of longitude will get a view. Only the eastern coast of North America will see the eclipse totality, as the Moon rises.

August 28, 2007:
Greatest eclipse is at about 10:40 and therefore visible across the Pacific region, from the Americas across to China. Throughout North America totality may be seen, although the Moon will set before that phase is completed for viewers on the East Coast.

February 21, 2008:
This event is centered on 03:25, making it yet another western Atlantic eclipse. All longitudes from the Middle East to the Pacific Ocean side of North America provide viewing locations for the entire eclipse.

After the above, there are no more total lunar eclipses until December 21, 2010. Although that one is visible in its entirety from throughout North America, you would need to go elsewhere to see the two in 2011. Following that there are no opportunities until 2014.

Obviously total lunar eclipses make themselves available more often than their solar equivalents, but they are not so frequent that there is a huge number occurring within your lifetime. If you live in North America, you have two opportunities in 2003, and then another later in 2004. After that there's a wait until 2007. Take the chance while you can. In their own way, lunar eclipses are fascinating.

16

An Eclipse Whodunit

He who doubts from what he sees
Will ne'er believe, do what you please.
If the sun and moon should doubt
They'd immediately go out.

William Blake, *Auguries of Innocence*

In this book we have examined both the astronomy and the history of eclipses and seen along the way that these natural phenomena occasionally have represented pivotal junctures in the development of civilization. The interpretation of eclipses affected the outcome of strategic battles in ancient times, and even today these celestial events are regarded superstitiously by many. Witness the way in which they are noted in newspaper astrology columns, and how the lunar and solar eclipses in January and June 2001 prompted civil unrest in Africa.

A scientific (rather than superstitious) argument over a specific eclipse in medieval times so far has not been mentioned. This was the total solar eclipse that passed over parts of the British Isles shortly before the Synod of Whitby in A.D. 664. The major outcome of that great synod, which turned largely on a debate related to that eclipse, was that seven previously warring fiefdoms became united to form what we would now recognize as the nation of

England. If it were not for that, then the history of the world would be quite different, and this book would have been written, if at all, in some alternative language.

It is clear, then, that the eclipse of 664 was important. The peculiar thing is that the circumstances of the eclipse were misrepresented, both at the synod and in later accounts of what took place. If this dishonest dealing had not occurred then the outcome would likely have been rather different. Just how this was done is not yet clear, providing us with an eclipse whodunit at least the equal of most detective novels. Just who was responsible for this subterfuge at the synod, frustrating the opposition and, in consequence, altering the course of history?

THE MYSTERY OF WHITBY

The picturesque little fishing port of Whitby stands on the northeast coast of England, about 200 miles due north of London. Whitby owes its historical significance largely to the ruined abbey that stood there for centuries, having been founded in A.D. 658. Six years later, with the Abbess Hilda overseeing the hospitality for her guests, a great synod was hosted in Whitby, a meeting of ecclesiastical authorities that was in effect to decide the future of the Christian Church throughout the British Isles.

The Synod of Whitby in A.D. 664 has, over the intervening 13 centuries, achieved not only considerable significance in Church history, but also a popular reputation as a mysterious affair. For example, in *Absolution by Murder: A Sister Fidelma Mystery*, author Peter Tremayne sets his fictional thriller against the factual backdrop of the synod. As the jacket blurb for this 1994 novel explains, "When the Abbess Etain, a leading speaker for the Celtic

Church, is found murdered suspicion inevitably rests on the Roman faction." No such murder actually occurred (and Hilda was the only woman recorded as being directly involved in the synod), but the circumstances invoked by Tremayne are realistic, there being some considerable dispute between those two opposing parties.

Although no murder took place, astronomical truth was assassinated at Whitby, for which there is more than mere circumstantial evidence to suspect the Roman faction. In short, there is a real-life mystery about what transpired at the Synod of Whitby; the full truth still needs to be teased out from the various clues through the eons. In 664 the Roman party carried out a religious sting, fooling the King of Northumbria into rejecting the Celtic overtures and transferring his allegiance to the Roman Church.

EARLY CHRISTIANITY IN THE BRITISH ISLES

The present name of Whitby attests to links with lands across the North Sea. In Scandinavia many place-names end with "-by" (pronounced *bee* not *bye*), meaning "village," in the same way that other English locations often end with "-ton," a diminutive of "town." Just to the south of Whitby lie Scalby and Newby, the meaning of the latter name in particular being obvious. Back in the seventh century, before the Viking marauders began to arrive, Whitby was known as *Streanaeshalh*. That's an unfamiliar mouthful, so let us keep to the later name of the town and monastery. Rather than looking over the North Sea, the Whitby connection with which we are concerned here causes us to cast our gaze to the Mediterranean and the warmer climes of Italy, Greece, and Turkey. It is from those regions (using their modern names) that Christianity diffused to the British Isles.

Although Julius Caesar had twice ventured into southeastern England, the Roman conquest did not occur until about a hundred years later. Then, the emperor Claudius pacified most of the region we now call England, pushing the Celts, Picts, and other races back into the far corners of the isles, into Scotland, Wales, Cornwall, and Hibernia (Ireland). Trouble persisted with those peoples, eventually resulting in such steps as the construction of Hadrian's Wall.

In these early centuries of the Roman occupation, the state religion was pagan, based upon various deities associated with the planets (which term included the Sun and the Moon in that era). For example, Mithraism lauded the god of the Sun as being supreme, and the winter solstice—when the Sun stands still in terms of the rising point on its annual oscillatory path up and down the eastern horizon—was the major celebration each year, the festival of the unconquered Sun. The date of Christmas was derived by transfer from that pagan feast to the Christian holiday (literally, holy day) soon after the emperor Constantine the Great became sympathetic to Christianity in the early part of the fourth century.

As Christianity became the official state religion of the Roman Empire during the fourth century, missionaries spread to Britain and began converting the indigents to that faith. At the same time there was a gradual withdrawal after A.D. 350 of Roman influence from Britain, the final tie being cut in about 410. The cause for this withdrawal can also be traced back to the long-dead Constantine. He had shifted the imperial headquarters from Rome to Byzantium, causing its name to be changed to Constantinople, as it remained until the city's name was changed to Istanbul in 1930. With the seat of power removed to the east, the Italian peninsula was left largely unguarded as the Gothic hordes swept west from their homelands at the periphery of the Black Sea. The people still in Rome pleaded for assistance, but their cries fell on the deaf ears of those now largely Hellenized, living happily in Constantinople, and lording it over the Eastern Empire. As a result the Western Empire finally collapsed by 476, Rome having been sacked several times by barbarian tribes during that century.

In consequence Britain was no longer part of the domain of the Roman *State*. Affairs in those islands, however, could still be influenced by the Roman *Church*. That Church had other problems to deal with, such as surviving within a city (Rome) and country (what we now call Italy) occupied successively by various non-Christian rulers such as the Goths and the Huns. With the disintegration of order in Britain, the peoples known as the Jutes, Angles, and Saxons invaded from mainland Europe, and anarchy reigned as the Picts and the Scotti rampaged down from the north and the west. In Britain, as well as elsewhere in the west of Europe, famine and disease were rampant, and town life collapsed as the society previously organized as part of the Western Roman Empire simply disintegrated.

That society had previously produced a sufficient economic excess to support scholars, and in addition for the armies to hold back would-be invaders. With the collapse of the Empire, from the late fifth century Europe entered the Dark Ages. That term is commonly applied so as to reflect our scant knowledge of what happened during that era, in the absence of records kept by the learned men who were earlier employed as part of the bureaucratic system. One of the few places anywhere that scholars could work and maintain written records was in the Church and its associated monasteries.

Christianity had hitherto made little progress in the British

Isles, although there were outcrops of believers here and there. One important concentration was in Ireland, where missionaries from Crete and other parts of the eastern Mediterranean had arrived by the early fifth century. We will see later that the origin of those missionaries is significant. This early arrival of Christianity in Ireland led to the establishment of several monasteries, and Irish annals that show records of phenomena such as eclipses, comets, auroras, and volcanic clouds (from eruptions in Iceland) exist, dating back as far as A.D. 442. The Church in Ireland, totally disconnected from the Holy See in Rome, seems to have thrived in this era. Many of the bellicose peoples who had been penned into Ireland rebounded into Britain with the fall of the Roman dominion, and at length church missionaries followed eastwards across the Irish Sea in their wake.

One of these missionaries was Saint Columba (521–597), who traveled to Scotland. The organization that made its presence felt in the north of Britain at this stage is nowadays referred to as the Irish, Celtic, or Columban Church. Various monasteries were founded in northern Britain, and two of the most preeminent centers were those on the tiny island of Iona (in the Inner Hebrides, off the western coast of Scotland) and on another island called Lindisfarne (just off the coast of Northumbria, the northeastern part of England). The locations of these monasteries are shown in the map in Figure 16-1. In remote places such as these the torch of learning was carried forward through the Dark Ages.

To the south, in England, heathenism still reigned in the various kingdoms ruled by the Jutes, Angles, and Saxons, and pagan gods were worshipped. This began to change from about A.D. 597 when the first Roman Church missionaries arrived in Kent in the southeastern corner of England a short maritime hop from conti-

FIGURE 16-1. This map indicates the ground track across the northern parts of the British Isles of the total solar eclipse that occurred on the first day of May in A.D. 664. Several early ecclesiastical centers and monasteries are also shown.

nental Europe. At that time the Roman Church, under Pope Gregory the Great, was enjoying some stability after the Eastern Empire under its leader Justinian had reasserted itself in Italy from the middle of the sixth century. The Church had started a concerted effort to spread Christianity through Germany and the countries along the Danube, and Saint Augustine was sent west in an attempt to convert the heathens then occupying most of Britain. Augustine established the cathedral at Canterbury and set about his task, although he did not get far because he died in 604, but the changes he initiated were important.

Augustine converted the Jutes of Kent to the Roman Church, and served as the first Archbishop of Canterbury. The next targets for his successors were the constantly warring Angles and Saxons of the other six kingdoms within England (the Essex, Wessex, and Sussex of the Saxons, and the more-northerly East Anglia, Mercia, and Northumberland of the Angles). There was a problem, though, in that the Celtic Church was likewise trying to assert itself, using traditions and practices somewhat different from those of the Roman Church. One of the disputed matters might seem absurd to us now: it was the form of the tonsure, the way in which monks shaved the tops of their heads. The great fighting-ground, though, was the subject of when Easter should be celebrated each year.

THE SIGNIFICANCE OF EASTER

One of the great sources of schism in the early Christian churches was argument over the calculation of the date of Easter. This is called the *computus*. In principle its statement in the present epoch is easy: in any year Easter is the first Sunday after the full moon occurring next after the spring equinox.

The much-misunderstood problem is that the full moon and the equinox referred to in that statement are *not* defined by the Moon and the Sun in the sky, but rather by theoretical constructs invented for ecclesiastical usage. The "equinox" for Church purposes is stipulated to be the whole of March 21, whereas the astronomically defined instant of the equinox—when the Sun crosses the celestial equator—varies over a 53-hour range from March 19 to 21. The "ecclesiastical moon" is an imaginary body that is assumed to follow the 19-year Metonic cycle containing 235 lunar months (as discussed in Chapter 2), but with eight separate corrections, each of a single day interposed over a 2,500-year grand cycle. The "ecclesiastical sun" is likewise corrected with three single day jumps in a 400-year cycle: this is why a century year is not a leap year unless it is divisible by 400 (as was the case for the year 2000).

To give an example of how misleading the verbal statement above might be, astronomical full moon could occur on March 20 and the equinox on March 19, but still the ecclesiastical rules delay Easter Sunday by a month because the equinox is assumed to be March 21. Equally well, sometimes Easter occurs on the day when the Moon is full in the sky, even though the verbal statement seems to prohibit such an event.

The above is the contemporary position, stemming from the reform of the calendar promulgated by Pope Gregory XIII in 1582. That reform resulted in an Easter computus now used throughout the Western (Catholic and Protestant) churches, but not by most of the Eastern Orthodox churches. The latter persist in using the earlier Julian calendar and different rules, meaning that their Easter may agree with that of the West but equally well may be one, four, or five weeks later in many years. That is exasperating enough in itself, but if we step back to the first centuries of Christianity we find that the situation was even more confused.

At the time Constantine transferred his sympathies to Christianity in A.D. 312 there was a wide range of Easter practices being employed. This was because of both the slow communications in those days, and the lack of basic agreement between the various factions of the early Church spread around the Mediterranean and

Middle East. Various ecumenical councils were convened where the bishops from different regions met and discussed liturgical and doctrinal matters. These culminated in the Council of Nicaea in A.D. 325, held in a town now called Iznik, across the Bosporus about 60 miles southeast of Constantinople. The Nicene Fathers— the 319 bishops who attended—drew up the Nicene Creed, the formal statement of the underlying tenets of the Christian faith.

It has been much misstated that the Nicene Fathers laid down the rules for Easter, this construed as fact even in present-day papal missives. In reality all they did was this:

1. They agreed with an earlier council that all Christendom should celebrate Easter on the same day (an ideal that has never been achieved).

2. They made a statement that is covertly anti-Semitic (the major concern was avoiding the Jewish Passover, for reasons of self-identification similar to having the Christian Sabbath on Sunday, rather than Saturday as do those of the Judaic faith).

3. They referred the actual computation of Easter to the Church of Alexandria, in deference to the long Egyptian tradition of calendrical calculations based on celestial observations (for example, when Julius Caesar reformed the calendar in 46 B.C. he did so under the advice of an Egyptian, Sosigenes).

Despite this step forward by the Nicene Fathers there was still no agreement as to how Easter should be calculated. For two centuries the Roman Church refused to accept the dates stipulated by the Alexandrine Church. Both churches used the 19-year Metonic cycle, but while the Alexandrians assumed that full moon for ecclesiastical purposes should be taken to be the fifteenth day after new moon, the Romans insisted on the sixteenth.

Elsewhere other churches used their own schemes, producing their own Easter (or Paschal) tables that would be distributed throughout their dioceses to show when the various feasts should be held for some decades into the future. In particular some used an 84-year cycle, consisting of 4 Metonic cycles plus an 8-year addition. The latter stems from an ancient Greek invention called the *octaeteris*, whereby rather than having a single leap-year day every four years, the months instead followed the Moon with three extra of those lunar months being inserted into an eight-year cycle. The first Olympics starting in the eighth century B.C. followed this cycle, alternating between gaps of 49 and 50 lunar months rather than the quadrennial system we adopted for the modern Olympics.

It was this 84-year system that the first Greek missionaries brought to Ireland around A.D. 400, and was then employed by the Celtic Church. Not only was the cycle different, but also the rules allowed Easter Sunday to fall on the fourteenth day after new moon, making coincidence with Passover a possibility. Avoidance of such a perceived abomination had been the major concern of the Nicene Fathers.

The early Celtic Church, however, was disconnected from the Roman Church. After many quarrels with his Alexandrine counterpart, in A.D. 525 the Roman Pontiff asked Dionysius Exiguus, a learned monk from southwestern Russia who lived and worked in a monastery in Rome, to consider the Easter question and draw up Paschal tables for the next several decades. This Dionysius did, in fact largely adopting the *Alexandrine* full moon rule, but with a 19-year cycle.

It is from Dionysius's calculations that the erroneous year count of the Christian Era (the Anno Domini system) was later derived. He computed Easter dates forward for 95 years (5 cycles), from A.D. 532 to 626 inclusive, as the pope had requested. Dionysius also back-reckoned for 28 cycles (that is, the 4-year leap cycle of the Julian calendar multiplied by the 7-days-a-week cycle), each of 19 years, making 532 years in all. This took him back to the year we call A.D. 1. Let us look Dionysius's chronology, to see how it all came about. According to his thinking, the year A.D. 1 began with the circumcision of Jesus (when Jewish boys are considered to begin their lives). The Nativity was then eight days earlier (circumcision on the eighth day for Jewish male babies is stipulated in the Bible), on the traditional date of the winter solstice, December 25. Given the human gestation period it follows that the Incarnation, or Annunciation, when the Angel Gabriel told Mary that she would bear the Son of God, was nine months earlier, on March 25. That is the traditional (but not astronomically accurate, even at that time) date of the spring equinox. Dionysius reckoned the years for his Easter table, a count with each labeled the *Anno ab Incarnatione*, from March 25, 1 B.C. He seems to have been wrong by about four years, mistakenly thinking that the reign of Augustus Caesar should be counted from the year 27 B.C. This was when Augustus adopted that name, rather than 31 B.C. when, under the name Octavian, he became the *de facto* emperor by defeating Mark Antony and Cleopatra at the Battle of Actium. Dionysius then misinterpreted a biblical statement that Jesus was born in the 28th year of the reign of Augustus, resulting in the four-year error that has persisted ever since.

Leaving that digression aside, the important point is that the Roman Church used Easter tables based upon a continuation of Dionysius's computus from A.D. 532 right through until the Gregorian reform of 1582. When Saint Augustine arrived in Kent at the end of the sixth century, he brought with him Easter tables that were copied for spreading throughout the expanding domain of that Church, and continued over further 19-year cycles after Dionysius's own calculations expired in 626. The distant Celtic Church on the other hand maintained the 84-year cycle it had inherited from Greek sources much earlier.

A simple but important point must be made here. In this era reference to the "Roman Church" should be differentiated from the present-day "Roman Catholic Church." Although the latter derives from the former, here we are discussing affairs almost a millennium before the Reformation. (That was when the various Protestant churches split off from the Catholic Church: in England as a result of King Henry VIII's disputes with the Pope and in Germany through Martin Luther nailing his list of complaints to the church door.) By the term "Roman Church" reference is made to what might also be called the "Western Church" in contradistinction to the traditions through which the present Eastern Orthodox churches came about.

THE SITUATION IN NORTHERN BRITAIN

After the withdrawal of Roman governance and the incursions by numerous barbarian bands, various pagan religions were followed in England. Bordering on Scotland, Northumbria under King Oswald had become an adherent to the Celtic Church from about 633 onwards. In 642, though, King Penda of the more-southerly Mercians defeated Oswald in battle, and promptly dismembered him. The latter's followers collected various parts of his body and distributed them to several churches (his head went to Lindisfarne and is now in Durham), leading to a cult of Oswald and eventually to his sainthood. Heathenism dominated in Northumberland until Penda died in 655.

A new king, named Oswy or Oswiu, then seized the throne, and the region reverted to Christianity, in particular the Celtic Church. To the south, Mercia also became Christian. Previously there had been a pagan buffer zone between the spread of Celtic influence in Scotland and that of the Roman Church much further south, but now that buffer was gone and internecine confrontation was inevitable as each church vied with the other to spread its influence.

Heeding the Nicene Creed, the Celtic Church made some attempt to understand the doctrine of the Roman Church and sent at least two delegates to Rome to obtain information and instruction. These men were rich Northumbrian nobles; dictionaries of the saints recall them as Saint Benedict Biscop (628–689) and Saint Wilfrid (633–709). Benedict Biscop departed first in 653 (he made five visits in all), accompanied on that trip by Wilfrid who, after a protracted stay in France, returned in 658. After much discussion with the Pope on doctrinal questions, they were convinced of the rectitude of the Roman computus and returned with many valuable ecclesiastical items such silken cassocks and extensive collections of Church documents, including Easter tables calculated according to the rules adopted by Dionysius.

With Benedict Biscop and Wilfrid having been won over to the opposite side, much argument ensued on the Easter question, although there were other grounds of debate, such as the form of the tonsure as mentioned earlier and also the role and power of the bishops.

These matters were brought to a head in the 660s, apparently due to the fact that Oswy was married to a Kentish lady, Queen

Eanfleda, whose personal priest ensured that she followed the Roman rules for Easter. In one year it happened that the fourteenth day of the month counted from new moon was a Sunday and so the Celtic Church scheduled Easter for that day, whereas the Roman computus put it a week later because the fifteenth was the earliest day permissible under its rules. This meant that the Easter Sunday feasting of the Celts coincided with the Palm Sunday of the Roman Church, a day of atonement, resulting in King Oswy attending the festivities while his consort was fasting and so unable to join him.

The upshot of this was that Oswy decided that the matter must be brought to a resolution, and so he called the Synod of Whitby in 664. Our surviving accounts present what happened at the synod as being a triumph of reason over an inferior computus (but then the winner always gets to write the history). It seems that Wilfrid championed the cause of the Easter calculation of Dionysius, and the argument finally swung that way when he claimed the authority of Saint Peter, which much impressed Oswy, since he did not want to offend the keeper of the keys to the gates of Heaven.

The outcome was that the Roman Easter system became accepted throughout much of the previous Celtic domain, although the monks of Iona in the Western Isles held out on the 84-year cycle until 715. In the region we now call England, the Roman Church held sway throughout. Between 669 and 690 Theodore of Tarsus, Archbishop of Canterbury, was instrumental in bringing together the seven feudal kingdoms, united now in religion, to form what became the English nation.

The Synod of Whitby was a pivotal event then in both the history of the British Isles and the evolution of the calendar. It was the acceptance and preservation in Britain of the Easter tables invented by Dionysius Exiguus—while most of Europe was in chaos during the Dark Ages—that led to the dating system we use today. Because of this, it is important that we understand what took place at the synod.

The simple and conventional account given above is not the whole story of the synod, and scholars have suspected for some centuries that there was more to the transactions than first meets the eye. Just in the past few years some more light has been cast on the happenings through astronomical investigations. It seems that the Roman party accomplished a sneaky but successful subterfuge.

THE ECLIPSE OF A.D. 664

It might be imagined that King Oswy must have been a good and pious man, who had recognized the problem with the differing Easter dates through a domestic issue, had called experts together to discuss the matter, and had then made a wise decision based upon the arguments presented. There are several misconceptions there that need to be demolished.

As a matter of fact Oswy was a bloodstained monarch who had carried out many unchristian acts, including the murder of his cousin Oswin. Oswy had then founded various abbeys in the north of England not so much out of goodwill, but more as an act of expiation. There were already monasteries at York, Ripon, Lastingham, and Lindisfarne. In association with the last of those Oswy had built new establishments at Hartlepool and Gilling, plus Whitby as has already been mentioned, and another dozen abbeys in the region, all long-since lost in the mists of time. Each was administered as part of the Celtic Church until the synod led to their transfer to the Roman tradition.

This was a major transitory step and following the synod the delegates of the Celtic Church, abashed and defeated, hastily beat a retreat to Iona, after a 30-year ascendancy in Northumbria. As aforementioned there is an extensive account of the actual debate at Whitby that has been handed down to us, but again we need to remember that the victor writes the history, and so we might perhaps look for other definitive evidence of the circumstances, such as astronomical clues.

One thing that is not known for sure is the date of the synod, which seems remarkable: if it were so pivotal, why did no one mark down precisely when it occurred? Some historians have even argued that it was held in 663 rather than 664, but on a mistaken basis. We know that it was during the latter year, and the recent recognition of the significance of that year has led to the possibility of a good guess at the date being made.

Another great event occurred in the British Isles in A.D. 664: a total eclipse of the Sun. This is the earliest such eclipse to have been definitely recorded in England, the path of totality also crossing the northern parts of Ireland (Figure 16-1). As a recent research paper by Dublin academics Daniel McCarthy and Aidan Breen points out, it seems remarkable that the possible link between the synod and the eclipse had not previously been examined because the events occurred close in time and both involved the Moon.

Were the records perhaps fudged deliberately to obscure the connection? It seems certain that the Synod of Whitby was held at some time towards the middle of 664, but on an indeterminate date. The non-recording of the date of the synod, blurring its association with the eclipse, may have been part of a plan designed to fool potential opponents of the Roman Church. This is a suggestion I will argue below.

We know when the eclipse occurred because we can compute such things with utmost accuracy, given other eclipse records that allow the deceleration in the Earth's spin rate over the past few millennia to be ascertained. It was on the first day of May. We know the track that the eclipse took across northern England, as shown in Figure 16-1. Whitby is close to the center, but most of the major monasteries in the north of England were also within the path of totality. As part of his act of expiation, Oswy had only recently established many of those monasteries and they all practiced according to the rites of the Celtic Church.

SIGNS OF THE APOCALYPSE

History also records another major event in England in A.D. 664: an outbreak of the bubonic plague, which seems to have come soon after the eclipse. Not only that, but there were other matters of concern. Frequent auroras had been observed, reflecting strong solar activity, also making sense of reports that the sky seemed like fire during the eclipse: extreme solar prominences would fit in with this picture. To people like Oswy, recently converted to Christianity and told by missionaries and the Bible what to expect as signs of the Last Days, it must have seemed that the Apocalypse was at hand, God displaying his anger and demanding that they should change their ways. Let us imagine how Oswy might have reacted.

What could have brought God's wrath down upon them? Oswy would have made hasty inquiries and found that the pitch darkness of the total eclipse occurred only within a band occupied by his own monastic establishments following the Celtic Church. At the southern limit to the path of totality was the ancient monastery of York. Along with Christian centers to the south, York had long since converted to the Roman Church. To Oswy, the message was clear: God was telling him that the Celtic Church was the wrong sect to follow.

The apparent sequence of terrifying eclipse, auroras, pestilence, and then the synod seems too unlikely to have occurred by chance. That is, in the atmosphere of dread following the eclipse, it appears that Oswy hurriedly called the Synod of Whitby in an effort to assuage the vengeance of God. Oswy must have thought that his own establishment of several new monasteries under the "false doctrine" of the Celtic Church had provoked divine anger. Under such circumstances the outcome of the synod would have been preordained, and the accounts of the proceedings largely a charade to provide a covering story.

Why was the synod held at Whitby, a brand-new abbey, rather than one of the older, established monasteries like Lindisfarne? Because Whitby was right at the center of the path of totality, perhaps singled out in Oswy's mind as a place indicated by God. Oswy's discomfort in this respect would have been heightened by the fact that his daughter had been installed at Whitby as a novice under the tutelage of the Abbess Hilda.

The role of the eclipse in this connection, the fact that it must have predated the synod, and its involvement in provoking Oswy's transfer of allegiance, are confirmed by a letter to King Oswy from Pope Vitalian in 665, in which the Pontiff wrote: ". . . we know how you have been converted to the true and Apostolic Faith by the shielding right hand of God." The conversion in question was not from heathenism to Christianity, but from the Celtic Church to that of Rome. The "shielding right hand of God" here is the Moon, which had obscured the Sun in a swathe passing across Oswy's new monasteries. The story is complete; but there is a fly in the ointment.

THE FALSIFIED RECORD

The link between the eclipse and Oswy's instigation of the Synod of Whitby seems clear. The subsequent history of Britain, and a wide variety of other matters, hinges upon his decision to switch allegiance to the Roman Church. Above, though, I wrote "it seems remarkable that the possible link between the synod and the eclipse had not previously been examined because the events occurred close in time and both involved the Moon." How did the synod involve the Moon? The answer here is simple: through the dependence of the Easter computus upon the lunar phase.

The central argument at the synod revolved around whether the 84-year cycle used by the Celtic Church or the 19-year cycle of the Roman side provided a better representation of the lunar brightness variation, coupled with the assumed full moon date (fourteenth or fifteenth day after new moon). It seems pretty obvious that the recent eclipse that had spawned the synod provided a rather concrete test. How did the opposing parties' lunar tables compare?

We are sure that the eclipse occurred on May 1, both from modern astronomical calculations and also accounts of it preserved in monastic annals from Ireland and elsewhere in mainland Europe. But the surviving *English* account has it on May 3. This could not have been a simple slip of the quill because the ancient dating system passed on from the era of the Roman republic was still in use. In that system there is no possible ambiguity, no chance of a simple mistake having been made. Some deliberate manipulation must have taken place.

The explanation for this erroneous English account seems to be that the Roman Easter table had a date for the new moon given as May 3, and that is what was recorded as the date of the eclipse after the fact despite it having actually occurred on May 1. A solar eclipse can only occur at conjunction, and the sighting of the new moon is typically not until 30 hours later (refer, for example, to Figure A-6 in the Appendix). But new moon can only be seen just after sunset: if it is not quite visible one evening, you must wait another full day until your next chance. (Similarly, if you miss an hourly train by 5 minutes, then you must wait another 55 minutes for the next departure.) A tabulated date for new moon on May 3 could therefore be consistent with the eclipse having occurred on the first day of that month. This is all known with hindsight and modern technical knowledge though. At the time it seems that the Roman party purported the eclipse to have occurred on May 3, in accord with their lunar tables.

The May 1/3 discrepancy has long been a puzzle to chronologists, having been pointed out at least as early as 1590. Under the circumstances it might not be too strong a statement to say that the record seems to have been falsified, and we would like to know how this came about. Who was responsible?

THE ROLE OF THE VENERABLE BEDE

Earlier a brief account was given of how Dionysius Exiguus developed a year numbering system, counting from March 25, 1 B.C., which was later taken up and developed into the era definition we use, the familiar Anno Domini scheme. The person whose actions led to this adoption was the Venerable Bede, a Northumbrian monk born a decade after the eclipse and synod of A.D. 664. It was also he who transmitted the false record to us although, as we will see, there were extenuating circumstances. We need to look at the interconnections between the characters in this story to learn more.

Saint Wilfrid we have already met, as one of the two men (along with Benedict Biscop) responsible for bringing back various documents, including Easter tables, to Northumbria from Rome, having been convinced by the Pope that the Easter computus set out by Dionysius was the method ordained by God. At the Synod of Whitby it was Wilfrid who was the main proponent of this winning cause. A protégé of Wilfrid was a man named Coelfrid, who at the time of the eclipse was a monk at Gilling, also within the path of totality. In 673 Biscop provided the wherewithal for the foundation of a new abbey at Monkwearmouth (or simply Wearmouth), and Coelfrid was seconded to assist, taking with him copies of annals recording what had occurred in 664, including the falsified date of the eclipse. In 681 Coelfrid moved up the coast a few miles to become the abbot of another monastery being built at Jarrow, and again took with him copies of the annals. From the age of seven, Bede lived with Biscop at Wearmouth, but then moved with Coelfrid to Jarrow, where he spent the rest of his life surrounded by the rich library of church documents collected by Wilfrid and Biscop.

Our knowledge of the early church history of England stems practically in its entirety from Bede's various accounts, written between 703 and 725 (he died in 735). Although his writings cannot be claimed to be perfect, he did a remarkably good job, resulting in "Venerable" usually being inserted before his name. Whenever he did find a mistake in some earlier record, he normally did his best to ferret out its cause and then put it to right. In the case of the eclipse, though, his account is quite peculiar. It is mentioned several times in his annals, and emphasized in various ways, Bede writing that it was an event "which our age remembers." The implication seems to be that he was distancing himself from the record and insinuating that there was something wrong with it.

Why, then, did he not correct it? Bede was quite capable of working out when the eclipse actually occurred. The answer seems to be that it was such a sensitive issue. Bede was writing only a few decades after the event, while the Celtic Church was still powerful in Scotland and Ireland and the hold of the Roman Church over England was tenuous. There were good reasons involved with church power and politics to cover the matter up then. Bede also had personal reasons: the misstatement of the eclipse date seems likely to have come from Wilfrid, who was still alive when Bede was first writing, and Wilfrid was a close colleague of Bede's spiritual father and mentor, Coelfrid.

It seems probable that Bede recognized that something was amiss, but did not feel able in his own lifetime to remedy it. Rather, he left a clear indication of the problem in the confident expectation that some later scholar would rectify matters. Perhaps the time has come.

THE FINGER POINTS AT SAINT WILFRID

What seems to have happened at the Synod of Whitby is that the Roman party was opportunistic. Wilfrid was arguing for the perfection of the Easter tables he had brought from Rome. The eclipse was startling and Wilfrid and his cohort wanted to play upon it, and so at the synod they bluffed that it had happened on May 3 and pointed to that date as having been predicted for new moon in their boasted tables. The synod was held at least several weeks after the eclipse and few people would have been able to recall for certain when the darkening of the Sun occurred. A modern court case drama might provide a good parallel to consider. When a lawyer asks a witness what they were doing at some specified lunchtime several weeks previously, an immediate definite response usually represents foreknowledge that the question was going to be asked and a checking of diaries. Alternatively, a suggestion from a lawyer to an unprepared witness that three Mondays ago he had a drink at a certain bar would likely evince a positive response if that witness habitually went there on that day of the week. It might well be that in the week in question he met his friends there for a game of pool only on the Wednesday, having been tied up with the laundry on the Monday, but details like that are quickly lost. Our memories are highly fallible. Many of us can recall where we were when we heard that President Kennedy had been assassinated or that the space shuttle Challenger had blown up—but what were the dates and the days of the week?

The Easter tables employed by the Roman Church were generally accurate enough for their desired purpose. At the Synod of Whitby, however, they were used as part of a deliberate subterfuge, a double-bluff deceit that led to the date of an important eclipse being incorrectly recorded and still causing puzzlement over 1,300 years later. May 3 was the eclipse date supplied to and recorded by Bede, two days later than it actually occurred, and he seems to have recognized that and was metaphorically waving a flag to ensure that the truth would eventually emerge.

It seems unfair to leave this shadow hanging over Saint Wilfrid and his party, because we cannot be sure that a deliberate misrepresentation of the facts occurred. On the other hand, the evidence seems strong. As is well known, ignorance of the law is no defense, whether it is the law of the land or the laws of celestial mechanics. An understanding of eclipse phenomena (such as how the Moon orbits the Earth while the Earth orbits the Sun) was still almost a millennium away, but simply recording correctly the date on which an eclipse occurred is hardly a highly technical problem. That the Easter table showed May 3 as the date of new moon and that Bede knew that any eclipse of the Sun would have preceded new moon is beyond dispute. History records various acrimonious disputes between Wilfrid and several kings and bishops; this seems to have been one fight that he managed to win, although in an underhand way, affecting us all in the end. If Wilfrid had not lied about the date of the eclipse of A.D. 664, the unfolding history of civilization would likely have been quite different and we might not be here to discuss it.

THE SIGNIFICANCE OF ECLIPSES REVISITED

The great English poet Thomas Hardy (1840–1928) began one of his verses as follows.

At a Lunar Eclipse
Thy shadow, Earth, from Pole to Central Sea,
Now steals along upon the Moon's meek shine
In even monochrome and curving line
Of imperturbable serenity.

He got the geometry right—the "curving line" of the terrestrial shadow—but one wonders what he meant by "monochrome." People nowadays imply "black and white" by that term, the usage postdating the invention of television, but the lines were written in 1903. Strictly the meaning of monochrome is "one color only," and with that intended meaning Hardy would be correct: the sole color is red.

This coloration has been recognized for eons. In the opening Chapter I mentioned the lunar eclipse that preceded the victory of Alexander the Great at the Battle of Gaugamela (or Arbela) in 331 B.C.; one account tells how the newly risen orb appeared: "But about the first watch the Moon in eclipse hid at first the brilliance of her heavenly body, then all her light was sullied and suffused with the hue of blood." When next you see a lunar eclipse, imagine Alexander rallying his troops, urging them on, telling them with assuredness how they will conquer the Persians after being blessed with this sign. He convinced them that it augured well for their endeavors, and their futures.

Of such human foibles and barbarity, Thomas Hardy despaired:

How shall I like such sun-cast symmetry
With the torn troubled form I know as thine,
That profile, placid as a brow divine,
With continents of moil and misery?

Let me close with one of the most famous eclipses of antiquity, about which the arguments continue. We met it in Chapter 3. It remains a notable episode, a prime example of how eclipses have affected the affairs of humankind. Thales may have guessed that a solar eclipse was due in 585 B.C., but one doubts whether he predicted its date and location, the history being invented after the fact. Herodotus, writing more than a century later, gave this account:

> ... there was war between the Lydians and the Medes for five years ...They were still warring with equal success, when it chanced, at an encounter which happened in the sixth year, that during the battle the day turned to night. Thales of Miletus had foretold this loss of daylight to the Ionians, fixing it within the year in which the change did indeed happen. So when the Lydians and Medes saw the day turned to night, they ceased from fighting, and both were the more zealous to make peace."

Whether promoting peace or provoking renewed fighting, without eclipses history would have been quite different, and this book would never have been written, or read.

Appendix:
Calculating Eclipses

We have seen in the main text how various forms of eclipse result from cosmic alignments and found evidence of the regularities in their occurrences. There are, though, many details that we glossed over, postponing their discussion to this Appendix. This was for two main reasons: one is that too much mathematical discussion tends to interrupt the flow of narrative, and the other is that many readers will feel uncomfortable with such analysis anyway.

In fact the calculations involved in eclipse prediction can be understood quite simply, once one has learned a little about how celestial objects move in their orbits. No higher math is needed than straightforward arithmetic, as you will see if you follow the arguments through as they are laid out below. In doing so you will gain a greater appreciation not only of eclipses, but also of our calendar, of how the movements of the Moon and Sun affect our climate here on the Earth, and various other matters. You will also catch a glimpse of how various ancient civilizations discovered the ways the eclipse cycles work, despite the fact that it would yet be many centuries before the nature of planetary orbits around the Sun was comprehended, in renaissance Europe. It is worth the little effort.

THE EARTH–MOON BINARY PLANET

Simple accounts of the Solar System often start by saying that the Earth orbits the Sun, and as it does so the Moon revolves around the Earth. While this is a reasonable first step, it is not quite true. Many stars are said to be *binary*: pairs of stellar bodies locked together in mutual gravitational embrace, each orbiting the center of mass of the duo. Similarly the Earth–Moon system can be thought of as being a *binary planet*.

The mass of the Moon is about one part in 81 that of the Earth. There are larger natural satellites elsewhere in the Solar System, such as Jupiter's Ganymede and Callisto, Saturn's Titan, and Neptune's Triton, but they are smaller in proportion to the mass of the associated planet. The only exception is Pluto and its moon Charon, discovered in 1978; Charon is about one-eighth the mass of Pluto, so that system certainly comprises a binary planet, although they are both tiny.

In the case of the Earth–Moon system, one should really say that the pair orbits their combined center of mass, which is termed the *barycenter*. In turn the barycenter orbits the Sun. The barycenter is on the line joining the middles of Earth and Moon, and the relevant calculation places it about 2,900 miles from the core of our planet. Because the terrestrial radius is about 3,964 miles, the barycenter is *within* the Earth, as shown in Figure A-1. As the Moon orbits, the Earth also swivels around this point, as indicated in the diagram.

Generally we are not aware of any wobble in our movement, but by the same token we tend not to notice that we are speeding along on our path around the Sun at near 18.5 miles per second (almost 67,000 mph). This velocity varies between about 18.2 miles per second in early July and 18.8 in January. Similarly, on

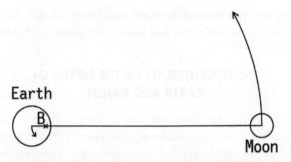

FIGURE A-1. The Earth and the Moon orbit their mutual center of mass, which is termed the *barycenter* (B). The barycenter happens to be within the Earth because our planet is so much bigger than the Moon. The two bodies are *not* here shown to scale compared with their separation.

the equator you are whizzing along at more than a thousand miles an hour, as you spin around the Earth's center. In our everyday frames of reference we are unaware of such movements. Wobble and change speed we certainly do, as we revolve around the Sun.

The Earth and the Moon rotate about the barycenter quite independently of the fact that they both spin on their central axes—the Earth once a day and the Moon, it happens, exactly once a month. The Moon therefore keeps basically the same face towards us at all times. We say that it is "tidally locked." Over the eons the lunar spin rate has been damped by Earth's gravity, because the Moon's mass distribution is not uniform. There is a greater density beneath the lunar nearside, displacing its center of mass away from its axis of symmetry, and the pull of the Earth keeps that greater mass directed towards us.

Finally, I wrote above that the barycenter is about 2,900 miles from the Earth's center, but actually its position varies. This is because the separation of the Earth and Moon changes, the lunar orbit being noncircular. We will learn more about this below.

THE ECCENTRICITY OF THE ORBITS OF EARTH AND MOON

The terrestrial orbit about the Sun is not a circle, its deviation from such a shape being defined by a quantity that astronomers call the *eccentricity*. The symbol used for this is e. A circle is defined as having $e = 0.0$ precisely, whereas the Earth has $e = 0.0167$ in the present epoch. Over many millennia this value changes and reaches a maximum value of almost 0.06 at times. This affects the climate because the influx of solar energy to our planet would then vary between perihelion and aphelion by a larger proportion than at present. The noncircularity of the orbit also causes the speed variation mentioned above. The effect is like a child on a playground swing, the highest velocity being achieved as the swing moves through the lowest point in the oscillation.

Currently we pass perihelion in early January and aphelion in early July (often on July 4, in fact). In consequence the Earth is moving slowest in July, during the warmest season in the Northern Hemisphere, soon after the summer solstice, and as a result summers in the north tend to be longer but cooler (the Sun being more distant) than those in the Southern Hemisphere. Similarly the winters in the north are shorter and milder than they would be otherwise. This will not persist forever because the dates of perihelion and aphelion advance by about one day every sixty years in the calendar we use. (That calendar was designed with a leap year scheme aimed at keeping the spring equinox on about the same date for ecclesiastical purposes, in particular the calculation of the date of Easter. If we wanted to keep perihelion and aphelion on the same dates instead then we would need to revise the calendar, and insert some *extra* leap years, rather than losing some as we do at present, as in 1800, 1900, 2100, 2200, and so on. Further matters concerning the calendar are discussed below.)

Now let's consider the Moon. The shape of the lunar orbit about the barycenter is likewise noncircular, having an eccentricity $e = 0.0549$. With an average separation of 238,850 miles, the lunar distance varies between about 225,740 miles at perigee and 251,970 miles at apogee, so long as that eccentricity is maintained. In fact, it is not. While the Moon is in a secure orbit (that is, it is gravitationally bound to the Earth), the attraction of the Sun perturbs its path in a cyclic fashion, and the lunar eccentricity varies fairly rapidly between 0.044 and 0.067.

This means that the barycenter moves rather erratically back and forth within the Earth, but let us lay that aside for simplicity, and in the following discussions and illustrations just imagine the Moon to orbit the center of the Earth. Keep in mind, though, the fact that effects like the motion of the barycenter are significant if one wants to compute accurate eclipse paths.

Figure A-2 shows the shape of the lunar orbit, compared to a circle. The Earth–Moon distance only changes by a small amount, but that is very significant with respect to the nature of eclipses. When the Earth is at its mean distance from the Sun, the solar orb has an apparent angular diameter of 0.533 degrees. That is the size of the light source that the Moon must entirely obscure to produce a total solar eclipse. Using the perigee distance of 225,740 miles mentioned above, with a diameter of 2,160 miles, the Moon subtends an angle of 0.548 degrees, and so is able to cover the Sun completely: a total eclipse. At apogee the lunar angular diameter is

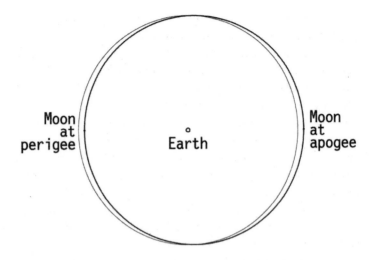

FIGURE A-2. The orbit of the Moon about the Earth (more strictly, about the barycenter) is not circular. Here the lighter curve is a circle centered on the Earth, while the heavier line is the elliptical lunar orbit. The Earth is shown to scale; the size of the Moon is equivalent only to about the width of the line depicting its path.

only 0.491 degrees, and so this time when the centers of Sun and Moon line up the latter cannot completely obscure the former, and so there must be a bright ring around its circumference: an annular eclipse. (The effect of the varying lunar distance was shown schematically in Figure 2.1.) Note that these angular sizes were calculated using the *average* eccentricity of the lunar orbit. The figures will change slightly as the eccentricity varies. For simplicity, in further discussions of the lunar orbit we will depict that orbit as circular, but remember that it is actually an ellipse.

Apart from the above we must also take into account the

noncircularity of the Earth's heliocentric orbit. This results in the apparent size of the Sun oscillating during the year, altering the target the Moon must obscure. The small eccentricity of the terrestrial orbit results in our separation diminishing to near 91.4 million miles at perihelion before growing to 94.5 million miles at aphelion, the apparent diameter of the Sun therefore changing between 0.542 and 0.524 degrees. Obviously this will also affect whether a solar eclipse is total or annular, if the Moon happens to be at a distance giving it an apparent size near those solar limits.

There are other complications. It was effectively assumed above that the potential observer is at the barycenter, which is not realistic of course, since it is deep underground! The size of the Earth is a significant fraction of the Earth–Moon separation, and so the angular size of the Moon someone will see depends to some extent upon his or her location on the surface of the planet. Imagine, for instance, that you are gazing at a full moon that has just risen above the eastern horizon at sunset. Six hours later, at midnight, you will be several thousand miles closer to it, and by sunrise you will have receded from the Moon again, all because of the Earth's rotation. This movement alters the angular dimension of the Moon by about a hundredth of a degree, and this may be critical when considering whether an eclipse seen from a certain location will be total or annular.

THE ORBIT OF THE EARTH AND THE CALENDAR

How long is a month? Even laying aside calendar months, with their variety of lengths (30 or 31 days, 28 for February but 29 in a leap year), the question is not a trivial one to answer. We start with a related question: how long is a year?

Before one can answer this, one must ask a simple but deceptive question. What is the crux of the matter at hand? In the case of the Gregorian reform—the alteration of the calendar by the Roman Catholic Church in 1582—the essential consideration was trying to maintain the date of the spring equinox. The "year" required for that aim is the time between such equinoxes, and that is *not* the same as the time taken to complete one orbit. In fact, due to several vagaries the notion of a period "to complete one orbit" has little meaning in itself. One must be very definite about the phenomenon of interest that is employed to define the start and end of the orbit, because different start and end points lead to different values for the year length.

Prior to the Gregorian reform, and after Julius Caesar introduced his eponymous calendar, a leap year had been employed every fourth year, producing an average year length of 365.2500 days. The Gregorian calendar reform amended the leap year rule such that the years A.D. divisible by 100 but *not* by 400 are common years (that is, not leap years), with no February 29. The result is that 97 leap year days are added to four centuries, and so the average year length is equal to 365.2425 days. (That comes from the fact that 97 divided by 400 equals 0.2425.) This "year"—the mean Gregorian year—is an artificial length of time, invented by humankind. One next needs to ask how long the natural or astronomical year might be, and compare the two.

The terrestrial orbit is shown schematically in Figure A-3. The large arrows indicate the spin axis of the Earth, which for the time being is assumed not to alter in orientation. Winter solstice occurs when that arrow is pointed as far as possible from the Sun, and at that time the Sun reaches its most southerly rising point during the year, on about December 22. In essence this is the

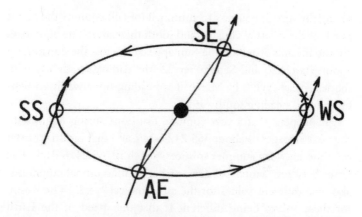

FIGURE A-3. The orbit of the Earth about the Sun (solid circle at center), in a slant angle view; in reality the terrestrial orbit is fairly close to circular. The positions of our planet at spring equinox (SE), summer solstice (SS), autumnal equinox (AE), and winter solstice (WS) are shown, the long arrows indicating the direction of our spin axis. The small cross indicates the position of the Earth when at perihelion (closest approach to the Sun) in early January.

shortest day (assuming you are in the Northern Hemisphere). The summer solstice around June 22 is when the Sun rises at its most northerly point, and the daytime hours are longest.

In between are the two equinoxes. Despite popular belief, it is not quite true that at the equinoxes the number of daylight hours equals that of nighttime hours, as the word "equinox" would suggest, because there is sunlight available for some time before sunrise and after sunset, plus other complicating factors. The equinoxes are defined astronomically, as follows. If one extrapolates the equator of the Earth out into the sky, the *celestial equator* is delineated as a circle cutting the celestial sphere into two. From spring (or vernal) equinox to autumnal (or fall) equinox the Sun is north of the celestial equator, and south thereafter. The equinoxes are the instants at which the Sun appears to cross the equator, on about March 20 and September 22 (the dates vary slightly with the leap-year cycle). In March it is heading northwards, in September it is heading southwards.

It happens that a year counted from one spring equinox to the next averages to about 365.2424 days, and that is distinct from the time between summer solstices, which is 365.2416 days. The times between winter solstices, or between autumnal equinoxes, also give different values for the astronomical "year." The reason for these values being different is that the speed of the Earth changes during its orbit. The average of the four is 365.2422 days, which is termed the *mean tropical year.*

It is a mistake, often made, to compare the mean duration of the year in the Gregorian calendar with the tropical year; the difference between them, about 0.0003 days, suggests that a single day correction might be required every three or four millennia. Actually the mean Gregorian year should be compared with the spring equinox year, the difference between these being but 0.0001 days, three times less. This might suggest that a correction of one day every ten millennia might be needed. However, the latter would again be based on a false premise: because the perihelion point of the Earth is moving, the lengths of all these "years" are changing from one century to the next. Another matter to consider is the fact that the Earth's spin rate is slowing, making the days longer, and so reducing the number of "days" in a year. It happens that, in the present epoch and continuing for a couple of thousand years, the Gregorian leap year rule actually provides for a better approximation to the necessary year length than most people imagine. Indeed many prominent astronomers have been led astray by misunderstanding what is going on.

The above should not be construed as a statement in praise of the Gregorian rule for leap years as used in the Western calendar. The system of dropping three leap-year days in four centuries results in the spring equinox shifting over a total span of 53 hours, between March 19 and March 21. In computing the date for Easter, the Church actually stipulates March 21 always to be the equinox, disregarding the phenomenon as defined astronomically. If in 1582 the Roman Catholic Church had really wanted to keep the equinox within a 24-hour period it could have done so by employing a 33-year cycle containing 8 leap years. This is because 8 divided by 33 equals 0.242424. . .(these two digits recurring). The average year length in such a scheme would be a little over 365.2424 days, closer to the desired spring equinox year than the Gregorian rule. More important, from an ecclesiastical standpoint, the briefer cycle time of 33 years would result in the equinox wandering by less than 24 hours.

In fact the Persian or Iranian calendar, which tries to regularize the date of the equinox for other cultural purposes, uses this 33-year leap cycle and so performs better than the Gregorian scheme in terms of astronomical accuracy. From the perspective of the Western calendar, which is used as the standard for commerce and communications throughout the developed world, because this is a secular calendar the wandering equinox, resulting from copying the Gregorian leap year cycle, is not of practical or symbolic importance. It is interesting to muse, however, on how our dating scheme might have been different.

There is, of course, an implication for eclipse cycle interpretation. We saw in Chapter 2 that the saros, the great cycle of eclipse repetition, leads to gaps of 18 years plus 10 or 11 days between eclipses of a similar nature. Whether that extra number of days is 10 or 11 depends upon the phasing against the leap-year cycle. To some extent that jitter would be ironed out if a 33-year calendar were used, as does Iran.

The saros is discussed in much greater detail below, but first we must consider other aspects of the apparent movements of the Sun and the Moon.

THE PRECESSION OF PERIHELION

Refer back now to Figure A-3. This diagram is a slanted view to allow the orientation of the Earth's spin axis to be clear, but even if it were drawn looking straight down from above the very low eccentricity would make the orbit's deviation from a circle difficult to identify. Note that a small cross is drawn on the terrestrial orbit. This cross indicates the position of the Earth at perihelion, equivalent to a date around January 4 in the current epoch, soon after the winter solstice on December 22. The perihelion point is slowly moving, however, owing to tugs imposed by the other planets, and this motion is called *precession of perihelion*.

The date of perihelion moves later by about one day every 60 years, so that 4,500 years into the future it will align with the spring equinox. About 750 years ago, perihelion and the winter solstice coincided. A full rotation of the perihelion point around the orbit takes about 21,000 years. These gradual alterations in relative alignment affect our climate, and are thought to be one of the causes of the Ice Ages. To demonstrate clearly what is meant by precession of perihelion, Figure A-4 depicts an imaginary precessing orbit with a large eccentricity. In the four-and-a-half orbits

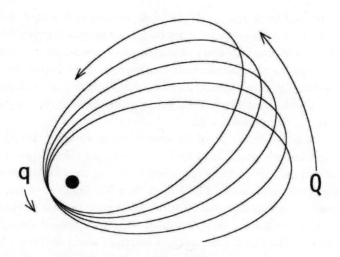

FIGURE A-4. Under the influence of various gravitational perturbations successive orbits precess (swivel around in their orientation) compared to the fixed stars. For clarity a highly eccentric (meaning noncircular) orbit is shown here. Both the perihelion point q and the aphelion point Q move counterclockwise from one orbit to the next in this diagram. Similarly both the terrestrial orbit about the Sun and the Moon's orbit about the Earth undergo precession in the counterclockwise direction.

displayed the perihelion point (labeled q) has turned through about 45 degrees, this movement being more obvious in the case of the aphelion point (Q).

The time taken for the Earth to return to perihelion, termed the *anomalistic year*, is 365.2596 days, almost one-hundredth of a day longer than 365 and a quarter. This might be considered the period to complete an orbit, but there are problems. If *that* year were used to design a calendar, then because it is longer than 365.25 days one would need not only a quadrennial leap year, but also an additional day every century, maybe a super-leap year with 367 days. If such a calendar were employed, predicated upon keeping the date of perihelion constant, then the dates of the equinoxes and the solstices would progressively move earlier in the year, and that would not do.

A better definition of the time taken to complete an orbit might be how long it takes the Earth to execute a 360-degree arc around the Sun. Because perihelion is moving counterclockwise, the Earth must traverse a little more than 360 degrees to reach it again. If one instead asked that the stars return to their previous positions in the sky, then the planet will have circuited through precisely 360 degrees, occupying a length of time called the *sidereal year*, which lasts for 365.2564 days. Again this is not really the sort of year wanted for setting up a calendar because the stars do not affect such things as our climate and seasons. The fundamental reference points we use are the equinoxes and solstices, but again those are not stationary, as we will see below.

THE PRECESSION OF THE EQUINOXES

The cyclic period of 21,000 years given in the previous section results from two quite different effects. One is the precession of perihelion as described: the gradual swiveling of the Earth's egg-shaped orbit. That length of time results from comparing the perihelion position with those of the equinoxes and solstices, but the latter positions are themselves moving, compared to a fixed reference frame based on the distant stars.

Imagine you are suspended in space far above the Solar System, looking down from the north. From this perspective you would see the perihelion point moving counterclockwise as in Figure A-4, the same direction as the orbits of the planets, and taking 110,000 years to complete a circuit. The equinoxes and solstices, on the other hand, would be seen to be moving in the opposite direction (clockwise) and taking about 25,800 years to turn. This gradual movement is called the *precession of the equinoxes*, and it has been a recognized phenomenon for more than two millennia, at least since the Greek astronomer Hipparchus described it in the second century B.C.; some historians claim that the Babylonians independently discovered the phenomenon centuries earlier.

The period of 21,000 years mentioned earlier results from these combined precessional effects, which are proceeding in opposite directions. You can check this with your calculator. Take the reciprocals of 25,800 and 110,000 (that is, 1/25,800 and 1/110,000), add them together, and take the reciprocal of the result. You will get 21,000 as the final answer. (The values are all approximate.)

If this is too complicated to visualize, the precession of the equinoxes may be better understood from Figure A-5. The long arrow represents the Earth's spin axis, pointing to the pole, P. The points marked A, B, C, and D are arrayed around the equator. When direction CD is aligned with the Sun it is the time of an equinox, and when AB is in the same plane as the direction of the Sun it is the time of a solstice. Although the spin axis remains in much the same orientation during one orbit, as depicted in Figure A-3, over millennia it gradually swivels, describing the clockwise circle shown at the top in Figure A-5. The movement is similar to the precession displayed by a toy gyroscope.

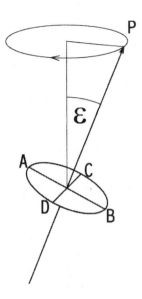

FIGURE A-5. The orientation of the terrestrial spin axis swivels around to complete a loop over a period of 25,800 years, measured against the distant stars. This is shown as the circle at the top of this diagram, P being the direction of the North Pole while A, B, C, and D are points on the equator. This motion is called the *precession of the equinoxes*. The angle denoted ε is called the *obliquity of the ecliptic*; in simple terms, it is the tilt of the Earth's spin axis.

The angle labeled ε is technically termed the *obliquity of the ecliptic*. It is simpler to think of it as being the tilt of the terrestrial spin axis, the angle of about 23.4 degrees between the line passing vertically through the plane of the terrestrial orbit (the ecliptic) and the line pointing towards the pole. The obliquity is therefore equal to the latitude of each of the tropics, because the lines marking the tropics are the extreme locations where the Sun passes overhead at the solstices. Various perturbations cause this angle to change slightly over millennia, but we will not worry about that here. I will merely note for interest that the slow, tiny decrease in the obliquity is of importance in astronomical interpretations of the first developments at Stonehenge, five millennia ago. At that time the obliquity was slightly bigger, and as a consequence the Sun rose on midsummer's day just a little further north than it does now. In testing such megalithic alignments one needs to take into account the value of the obliquity in the era in question.

THE CYCLES OF THE MOON

Imagine looking down upon the Moon's orbit (as in Figure A-2) from the depths of space, out among the stars. As with the sidereal year, one can define a *sidereal month* as the time the Moon takes to return to the same position relative to the stars; that is, to complete a 360-degree circuit around the Earth. This sidereal month lasts 27.32166 days, taking a long-term average value to smooth out short-term erratic variations.

The sidereal month is significant in that it is also the time the Moon takes to spin on its axis, so that it perennially points the same face towards us. Nevertheless we were able to map more than half of the lunar surface before satellites were launched to return images of the far side, because we can peek just beyond the eastern and western limbs of the Moon at different times, the lunar orbit not being circular. We can also see over the poles slightly, and overall 59 percent of the Moon can be mapped from Earth. Figure 1-2 shows these effects in action.

Is the sidereal month the one we need for eclipse computations? Well, not really. The reason is demonstrated in Figure A-6. It is the *synodic* month that is relevant for eclipses, as discussed in Chapter 2. This type of month describes the lunar brightness cycle, the length of time from one full moon to the next. Any particular synodic month may last for between 29.2 and 29.8 days, but the average taken over many years is 29.53059 days. That is a very significant figure in eclipse calculations.

OTHER TYPES OF MONTH

Just as the "year" comes in different flavors depending upon which precise phenomenon is of interest, there are other types of "month" beyond the two mentioned above. First we look at how the lunar orbit precesses.

The perturbations causing the precession of perihelion of the Earth's orbit are due to the other planets. Because these are mostly at great distances and all have much smaller masses than that of the Sun, the rate of precession is very slow: 110,000 years to complete a full turn. The Moon in its geocentric orbit is subject to much larger perturbations, because it is now the massive Sun that is mainly responsible for tweaking the lunar path. This results in its perigee precessing quite quickly, making a complete revolution in 8.85 years.

In consequence an alternative type of month can be defined, the *anomalistic month*, the time from one perigee to the next. This takes 27.55455 days on average, five and a half hours longer than a sidereal month. In Figure A-2 perigee was shown at far left; on the next orbit it would have moved counterclockwise by an angle of about three degrees.

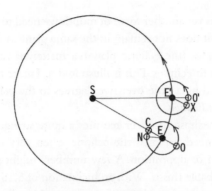

FIGURE A-6. The Earth (labeled E) orbits the Sun (S) while the Moon executes its own orbit about our planet. In the lower position, when the Moon is aligned with the Sun it is in *conjunction* (C), whereas when it is opposite (that is, 180 degrees from) that point it is at *opposition* (O). The Moon is said to be in *syzygy* when it is at either of these points. Eclipses can occur only at syzygy: a solar eclipse at conjunction, a lunar eclipse at opposition.

Although opposition is the time of full moon, conjunction is *not* the time of new moon. For the new moon to be seen it needs to have moved along its orbit to be sufficiently separated from the Sun in the sky such that it can be spotted near the western horizon just after sunset, and N is a typical new moon position. Conjunction may be thought of as being *dark of Moon*: our companion cannot be seen at all in the solar glare.

Now consider the second position of the Earth (labeled E'). When the Moon had turned 360 degrees about the Earth starting from O it reached position X, and the time taken to reach that point is a *sidereal month* (a month measured against the stars). To reach opposition again at point O' and produce the next full moon requires a little longer, a length of time called the *synodic month* (a month measured against the Sun). It is the synodic month of about 29.53 days over which the complete cycle of lunar phases is run, from dark of Moon at conjunction, to new moon an evening or two later, to first quarter, then full moon, then last quarter, and back again to conjunction. (Note that this diagram is not drawn to scale.)

There is yet another form of month we need to consider. The Moon's orbit does not remain in the same plane as that which the Earth occupies (the ecliptic plane), a matter of vast importance with regard to eclipses. This is illustrated in Figure A-7. The lunar orbit is tilted by a little over five degrees to the ecliptic, an angle called the *inclination*.

For an eclipse to occur requires a quite stringent alignment: the Moon needs to cross the ecliptic when very close to either conjunction or opposition. A few numbers might serve to show how improbable this is. With an inclination of 5.15 degrees and a geocentric separation of 238,850 miles, the Moon's distance above and below the ecliptic would oscillate between extremes of 21,440 miles, more than five times the terrestrial radius. In fact the Moon can deviate even more than this from the ecliptic because at apogee the geocentric distance is greater, and also its inclination varies between 4.96 and 5.32 degrees. Most of the time the Moon comes to syzygy (see Figure A-6) far above or below the ecliptic, and no eclipse occurs.

FIGURE A-7. The Moon has an orbit that is tilted slightly against the plane, called the ecliptic, in which the Earth circuits the Sun. In this diagram (*not* drawn to scale) we look sideways along the ecliptic, and note that the lunar orbit makes an angle of about five degrees to it, with an orientation that swivels around making alignments of all three bodies possible. An eclipse can only occur if the Moon happens to be crossing the ecliptic when at conjunction (producing a solar eclipse) or opposition (producing a lunar eclipse).

During each circuit of the Earth, the Moon crosses the ecliptic once travelling upwards, and once travelling downwards. These crossings are called the *nodes* of the orbit, the *ascending* and *descending* nodes respectively. Another way to define a month is using the interval between the Moon's nodal passages. This is usually called the *nodical month*, but another name often applied is the *draconic month*. The reason for the latter is that an eclipse can only occur when the Moon is passing a node, and ancient superstition said that a dragon then swallowed up the Sun (as in the story of Hsi and Ho related in Chapter 1), resulting in its obscuration; hence the term *draconic*. The nodical month has a mean duration of 27.21222 days. Unlike the anomalistic month, the nodical month is shorter than the sidereal month, and this requires an explanation.

Under various perturbational forces the nodes of all the objects in the Solar System are precessing. This is also true for the Moon. The Sun is causing the lunar perigee to shift, and similarly it is mainly responsible for the lunar nodes precessing. Actually, the nodes *regress* in that they move backwards (clockwise) around the orbit, in the same sense as the precession of the equinoxes, and this is why the nodical month is so short. This is illustrated in Figure A-8, both the precession of perigee and the regression of the nodes being represented, arrows indicating the sense in which they shift.

The above behavior of the Moon is of fundamental importance in the mechanism of eclipses. It is critical to follow what is going on. Figure A-9 shows the traverses of the Moon through three successive ascending nodes. Each time the Moon passes through that node the celestial longitude has been reduced by 1.44 degrees from the previous value (the origin of this step size is

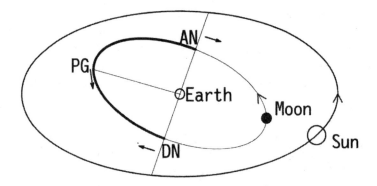

FIGURE A-8. How the lunar orbit precesses. To an observer on the Earth the Sun appears to orbit *us*, although the reality is that we are orbiting *it*. This produces the outermost path, which is close to being circular and restricted to the ecliptic. The lunar orbit is inclined to the ecliptic plane by just over five degrees, and so repeatedly crosses that plane at its ascending node (AN) and half a nodical month later at the descending node (DN). Half this time the Moon is above the ecliptic (heavy line) and half below (light line). The straight line connecting the nodes, passing through the Earth, swivels around in the clockwise direction as viewed from the north, as indicated by the arrows. This is called *regression*. A complete turn of the nodal line takes 18.61 years.

The perigee point (PG) precesses in the opposite direction, counterclockwise and in the plane of the lunar orbit. It takes only 8.85 years to complete a full turn.

In this diagram the Moon is shown near conjunction, but a solar eclipse could not occur because it is well below the ecliptic. For an eclipse to take place one of the two lunar nodes must be in the proximity of an imaginary line connecting Earth and Sun. *Total* solar eclipses occur when perigee is also near the node in question.

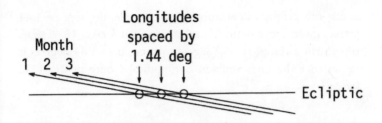

FIGURE A-9. The Moon ascends through its node (where its orbit crosses the ecliptic) once every nodical month, such a month having an average duration of 27.21222 days. Because this is less than the time it takes the Moon to complete a revolution about the Earth, the node moves clockwise along the ecliptic (from left to right in this view), successive values of the nodal longitude dropping by 1.44 degrees. The size of the Moon is shown to scale.

given below). As can be seen in Figure A-9, the shallow angle at which the Moon climbs up through the ecliptic results in some overlap with the path it took the previous month.

This implies that the Moon scans all of the sky along the ecliptic. Sooner or later the Moon at one of its nodes is certain to traverse the same longitude as the Sun, the latter being confined to the ecliptic. That's when a solar eclipse can occur: when a node occurs at conjunction. On the other hand, if the Moon reaches its node near a longitude 180 degrees from the Sun—at opposition, that is—a lunar eclipse will occur.

The lunar perigee precesses such that it takes about 8.85 years to progress around a full rotation. Similarly the lunar nodes perform a loop about the Earth, although in this case in the opposite (clockwise) direction, taking 6,798.3 days to do so. This regression period, about 18.61 years, is a fundamental cycle time that enters into eclipse calculations. For example, the step of 1.44 degrees given above stems from that period: if it takes 18.61 years to perform a complete 360 degree rotation, then 1.44 degrees is the distance the node shifts in one nodical month of 27.21222 days.

THE METONIC CYCLE

We have met with a variety of year and month lengths. From the perspective of calendar definition, we saw that the month of interest is the synodic month, the cycle time for the brightness phases of the Moon, currently lasting for an average of 29.53059 days (it is necessary to quote that to at least seven figures). Those readers with pocket calculators on hand may multiply that number by 235, for reasons that will soon be apparent, deriving a total of 6,939.69 days after rounding-off.

One could now argue about the proper length to use for a year, but the mean tropical year of 365.2422 days will do for these sums. If you multiply by 19 you get 6,939.60 days, rounded off. (In reality any particular set of 19 calendar years will contain either 6,939 or 6,940 days, depending upon whether five or only four leap years are counted among them.) It is immediately obvious that 235 synodic months last for almost exactly 19 years, the difference amounting to only 125 minutes. This 19-year period is called the Metonic cycle; we mentioned it earlier, in the Preface and in Chapter 2. There is more to be said about it, however.

The actual years we count in the Gregorian/Western calendar average to 365.2425 days, so that 19 of those will average to 6,939.6075 days. People often claim that the Gregorian calendar reform was necessary simply because the mean year in the Julian system (365.2500 days) was too long, and over the 16 centuries from Julius Caesar through to Pope Gregory XIII this resulted in the equinox arriving about 12 days too early. But that is only half the story.

From A.D. 532 the Metonic cycle had been employed in calculating the dates of Easter. For the cycle to be precise the average year length would need to be 365.2468 days (that is, 6,939.69 days divided by 19). Under the Julian calendar the average year lasted about 0.0032 days longer than this, and between 532 and 1582 these little differences had accumulated to exceed 3 days. In consequence the Moon in the sky was nowhere near the ecclesiastical moon followed by the Church tables for Easter, making Easter deviate substantially from full moon.

The Gregorian reform was therefore necessary to correct not only the Sun, but also the Moon, in terms of how closely the imaginary bodies encoded in the tables used to calculate Easter followed the movements of the real astronomical objects. The correction was designed to set those orbs right according to their parameters in A.D. 325, the time of the Council of Nicaea, when the fundamental tenets of the Christian faith were laid down.

Since 1582 the Catholic Church (joined later by many other Christian Churches) has continued to follow the Metonic cycle, but with two types of correction having been made. One is well known: the leap day corrections with three out of four century years being omitted and counted as common years instead, thus allowing the solar motion to be followed more accurately. But there is also a lunar correction, unrecognized by most people. This involves eight steps each of one day spread over 2,500 years. Using the figures cited above this correction appears to be near-perfect: 2,500 divided by 8 gives an average of once every 312.5 years, which is the same as the reciprocal of 0.0032 days (although more decimal places are really required in the calculations to be precise). Nevertheless it is a pretty good approximation to the real behavior of the Moon. (Note that many of the Eastern Orthodox Churches continue to follow the Julian calendar, so that their Easter is often on a different date.)

THE COINCIDENCES BETWEEN THE MONTHS

The Metonic cycle represents a coincidence between the synodic month and the solar year. There are three other definitions of the month we have met (the sidereal, anomalistic, and nodical months), each of them lasting for 27 days plus some fraction. In discussing eclipses we are not much worried about the stars, and so the sidereal month can be laid aside. But consider the mean lengths of the other three types of month:

Synodic month: $S = 29.53059$ days
(full moon to full moon)
Anomalistic month: $A = 27.55455$ days
(perigee to perigee)
Nodical month: $N = 27.21222$ days
(node to node)

Using those figures we can explore various matters of interest. For example, during a single lunation it is brightest at full moon, but not all full moons are equally bright: if opposition occurs near apogee then the full moon will be dimmer than during an opposition near perigee, because that orb is farther from us. We could ask then: How long is the period between those ultra-bright full moons near perigee? The answer is given by multiplying the synodic month by the anomalistic month and dividing by their difference $[(S \times A) / (S - A)]$, the result being about 412 days. That value is 13.94 times S: 13 complete synodic months plus about 94 percent of such a month, or 0.06 months (actually 1.64 days) short of the next full moon. Thus starting with a full moon at perigee, the fourteenth full moon will occur about a day and a half after perigee, and there will be a long-term cycle in full moon brightness.

One could take the broad question further. The brightness of full moon will depend upon how far above or below the ecliptic the Moon happens to be at opposition. One might imagine that brighter full moons occur when the Moon is at a node at opposition. In fact that would be the *dimmest* possible full moon, because that is when a lunar eclipse takes place. (Nevertheless, the brightest the Moon ever gets to be occurs just before a lunar eclipse, because then it is the nearest it ever comes to being precisely opposite the Sun in the sky, and that favors back-scattering of sunlight, plus the bonus of being closest if at perigee.)

Eclipses are what we are interested in here, and in this respect the month lengths we have labeled S, A, and N above have some remarkable relationships. We shall now examine just what sorts of cycles exist by doing a little numerical manipulation.

Full moon occurs near perigee about every 412 days, but over longer intervals there are cycles that are much more precise. Try doing the following sums on your calculator (the justification for them will soon become apparent):

$$223 \times S = 6{,}585.32 \text{ days}$$
$$242 \times N = 6{,}585.36 \text{ days}$$

This means that after 223 synodic months the Moon has returned to close to the same node as at the start of that sequence. The difference amounts to merely 51 minutes.

We are also interested in when perigee occurs, so consider the anomalistic month:

$$239 \times A = 6{,}585.54 \text{ days}$$

That is only about five hours longer than the canonical 223 synodic months above.

Shortly we will see the interval of 6,585.32 days to be extremely significant, but first we must learn about yet another type of year: the eclipse year.

THE ECLIPSE YEAR

In Chapter 2 we met a length of time termed the *eclipse year*. It was noted to last for about 346 days. Now we will see how it comes about.

A few pages back we saw that the time required for the lunar nodes to revolve once is 18.61 years. If the lunar orbit were stationary, in that the nodes were fixed, then the Sun would pass through those nodes once per solar year and we would get eclipses only on certain calendar dates. This is not the case, though. Because the nodes are regressing the Sun gets to them earlier, producing a type of year that is somewhat shorter than the solar or calendar year. Just how short may be calculated as follows.

Adding unity onto 18.61 to account for that revolution of the nodes, one derives a period equal to 18.61/19.61 times the solar year of 365.24 days (not worrying too much about the last decimal places), or 346.6 days, and this is the eclipse year. Pairs of solar eclipses are easily identified in tables, separated by about half an eclipse year, such as June 10 and December 4, 2002.

THE SAROS

The cycle of 6,585.32 days or 18.03 solar years is the saros, which we initially discussed in Chapter 2 without detailing its origin. This is the period over which conjunctions and oppositions repeat, making an eclipse possible.

Although the contrasting astronomical year lengths all involve fractions, each discrete calendar year must contain a whole number of days. Consider the saros counted off against our calendar. If there are 4 leap years within it, that cycle represents 18 years and 11 days, but just 10 days if by chance there are 5 leap years.

The saros contains close to, but not precisely, 19 eclipse years. Nineteen of *those* years persist for 6,585.78 days, which is 0.46 days longer than the saros. This means that when the syzygy passages repeat after 6,585.32 days, the lunar node is not quite in the same place as it was one saros earlier, because there is still 0.46 of a day to go. We can work out how much that equates to in terms of celestial longitude by expressing it as a fraction of the eclipse year and multiplying by 360 degrees; the answer is about 0.48 degrees, which is just less than the angular diameter of the Moon.

The situation can be visualized more easily by reference to Figure A-10. In Figure A-9 we were looking at successive nodal transits, producing a longitude jump of 1.44 degrees. Now we are

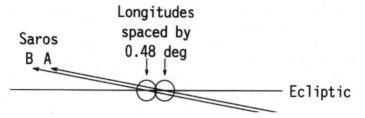

FIGURE A-10. After a complete saros the Moon comes back to a node just 0.48 degrees away from where it was 18.03 years before. Unlike in Figure A-9, the second saros (labeled B) starts with the longitude being enhanced (the node has moved counterclockwise, towards the left).

considering the situation after a whole saros. During that time the node has circuited the Earth 19 times, but returns to a position just 0.48 degrees from where it began the saros; the longitude is actually *enhanced* rather than *reduced* by that amount. As Figure A-10 shows, because the Moon is of slightly larger angular extent than this longitude jump, the lunar disks just overlap in terms of their positions from one saros to the next.

We know that the Sun is of virtually the same angular size as the Moon. Does this bare overlap mean that an eclipse occurring at the start of one saros will result in a miss at the start of the next?

REPEATING ECLIPSES

Regarding Figure A-10, one can see that although the lunar disk has shifted by 0.48 degrees in longitude, practically a whole diameter, because the Moon is crossing the ecliptic at such an oblique angle it will progressively cover most of the other disk before receding. Let us consider this in more detail.

Figure A-11 shows the trajectory that would give a grazing eclipse: the limb of the Moon just touches against the apparent edge of the Sun in the sky. Using quite simple geometry it is possible to calculate the value of the *ecliptic limit*, the longitude difference between the node and the Sun at which such a grazing eclipse would occur.

Actually there are distinct values for the ecliptic limit depending upon the specific conditions, because several varying parameters affect the calculations: the apparent sizes of both Sun and Moon depend upon our distances from those orbs, and also the inclination of the lunar orbit oscillates. Call the ecliptic limit L for shorthand purposes. Taking the most unfavorable values for the

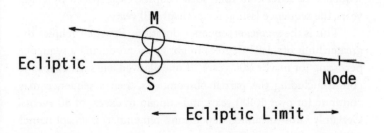

FIGURE A-11. A grazing eclipse of the Sun (S) would result in the situation depicted here. We know the angle at which the Moon (M) crosses the ecliptic at its node; this is the inclination, about 5.15 degrees. Then it is a simple geometrical matter to calculate the *ecliptic limit*, the maximum separation in longitude between the Sun and the node that will result in an eclipse of some stipulated type.

above parameters, if L is below 15.35 degrees then at least a partial solar eclipse is certain; if L is less than 9.92 degrees then a total or annular solar eclipse is certain. If L were below respective limits of 18.52 and 11.83 degrees, then such eclipses are possible but not certain.

The precise limits are not important. The significant factor to note is that they are all much greater than the step of 0.48 degrees that occurs from one saros to the next. This means that once the Moon gets into an orientation such that it passes a node within the ecliptic limits, for many following saronic cycles it will continue to produce eclipses.

Consider first the most stringent limit above, the range of 9.92 degrees certain to produce a total or annular eclipse. This is a permissible range for each side of the Sun, so that the total range in nodal longitude is almost 20 degrees. It will take 41 or 42 steps of 0.48 degrees to cross that distance, meaning that there will be a sequence of at least 40 total solar eclipses, each spaced by 18.03 years, the sequence lasting for perhaps 750 years.

This is the *minimum* sequence duration. For total eclipses the greater limit of 11.83 degrees might apply, producing a sequence persisting for maybe 900 years. If one allowed *any* solar eclipse to count, including the partial obscurations, then a sequence may continue for over 1,400 years and contain in excess of 80 events. Certainly saros (meaning "repetition," remember) is an apt name!

THE ECLIPSE SEASONS

The above does not mean that there are only 70 or 80 solar eclipses spread over 13 or 14 centuries, with gaps of almost two decades between them. Eclipses are much more common than that. During an eclipse year the Sun passes through the positions of both lunar nodes, and although the Moon may not be *at* its node, the ecliptic limits calculated above make it possible for solar eclipses to occur during the eclipse seasons that last while the Sun is traversing those limits.

The lengths of such seasons depend upon the eclipse type in question. Consider the widest, the ecliptic limit of 18.52 degrees making partial eclipses possible. The full longitude range is a little more than 37 degrees. Because the Sun moves through slightly less than a degree of longitude per day, the eclipse seasons are over 37 days long, but they slide through our calendar year, there being two such seasons (one for the ascending and one for the descending node) in each eclipse year.

Multiple eclipses can occur within an eclipse season: because a synodic month lasts less than an eclipse season, it is feasible that there will be two solar eclipses close together. In the year 2000 there were partial eclipses on July 1 and 31; these will repeat one saros later on July 13 and August 11, 2018, and again on July 23 and August 21, 2036. Such pairings of partial eclipses are possible because of the wide ecliptic limits; the narrower limits for total eclipses are not so generous.

THE ROLE OF THE METONIC CYCLE IN ECLIPSES

Eclipses recur in sequences separated by one saros, which lasts for 18.03 solar years (very close to 19 eclipse years). At any time there are many interleaved saronic cycles in action: 39 at present. Astronomers label these cycles with numbers. For example, the total solar eclipse of August 11, 1999 is part of saros 145, a sequence that began with an eclipse on January 4, 1639, and will end with the

77th on April 17, 3009. The next in this sequence is that cutting across the United States on August 21, 2017: book your viewing location now. Similarly the total solar eclipse of June 21, 2001 is part of saros 127, which consists of 82 eclipses between the years 991 and 2452.

(It may be noted that for the sake of clarity I have been a little lax in my usage of the term "saros." Correctly the word applies to the period of about 18.03 years after which eclipses repeat, whereas a phrase like "saros 145" refers to a whole sequence of eclipses spaced by such gaps. The intended meaning in each case should be clear enough.)

The saros is not the only cycle important in eclipse prediction. Earlier we met the Metonic cycle of 19 solar years and saw that it is of fundamental significance in calendar matters. After 19 years, 235 synodic months have elapsed, bringing the conjunctions and oppositions back to the same phase, to within a few hours. The Metonic cycle lasts for 6,939.6 days.

Break out the pocket calculator again. Multiplying the eclipse year (346.62 days to five figures) by 20 you will derive 6,932.4 days, which is just 7.2 days short of the Metonic cycle.

The implication of this is that after 19 years the Moon comes back to be not much more than seven degrees from its node, and another 19 years later it returns to a position again advanced by seven degrees. The maximal eclipse season we described above lasts while the Sun moves through 37 degrees, and the Moon's position may skip through that taking steps separated by seven degrees but 19 years apart. That is, there may be a short sequence of four or five (and just possibly six) eclipses separated by 19-year gaps, occurring at the same time of year.

The ecliptic limit chosen there is the largest possible, which is appropriate for partial eclipses. Regarding total and annular eclipses, three or four will occur in these brief sequences related to the Metonic cycle. For example, consider the total solar eclipse due on December 4, 2002. This will be followed by another such event on the same date in 2021. Looking back in time, there was an annular eclipse in 1983, and a partial eclipse in 1964, all on December 4. After 2021 the lunar node slips out the ecliptic limit, but re-enters on the other side a month earlier in the calendar with a partial solar eclipse on November 4, 2040, followed by three annular eclipses on similar dates in 2059, 2078, and 2097. These are all instances of the 19-year Metonic cycle gap, then.

THE MINUS 10 OR 11 DAY JUMP

The effect of the saros is that eclipses repeat on intervals of 18 years plus 10 or 11 days. But if you look at a tabulation of past eclipses you will find that there are sequences with interstitial periods of a year *minus* 10 or 11 days, with three or four eclipses in a row. For example:

February 15, 1961; February 5, 1962; January 25, 1963;
    January 14, 1964 (total, total, annular, partial eclipses of the Sun);
July 17, 1981; July 6, 1982; June 25,1983 (partial, total, partial eclipses of the Moon);
September 2, 1997; August 22, 1998; August 11, 1999;
    July 31, 2000 (partial, annular, total, partial eclipses of the Sun).

The reason for this is easy to see. Solar eclipses occur at conjunction, and conjunctions are spaced by synodic months; similarly for lunar eclipses at opposition. Twelve synodic months last for 354.37 days on average, which is 10.88 days short of a solar year.

On that basis one might expect eclipses to recur spaced by 354/355 days, but for how long could the sequence continue? The answer is given again by the lengths of the eclipse seasons, and the spacing between them. The longest eclipse season lasts for just over 37 days. The spacing of the eclipse season centers is equal to the eclipse year, 346.62 days, which is 18.62 days short of a solar year of close to 365.24 days. Therefore the eclipse seasons step backwards through the solar year in jumps of 18.62 days. At the same time the twelfth conjunction is stepping back by 10.88 days every year, producing a relative change of 18.62 - 10.88 = 7.74 days. Within a 37-day partial eclipse season one might get a sequence of a maximum of five solar eclipses in consecutive years (i.e., four steps of 7.74 days, equal to just below 31 days), usually less. Using the more stringent limits for total eclipses, a lower number of events appear in such a chain. These chains of eclipses just arrive consecutively 10 or 11 days earlier on the calendar (equivalent to 355/354 days *later*).

Turning to lunar eclipses, the ecliptic limits are more restricted, and as a result only pairs or trios with this spacing are identified. This is the reason for the patterns seen in Figures 15-5 and 15-6.

THE 3.8-YEAR GAP

The saros is a wonderful cycle: not only do eclipses recur with 18.03-year spacings, because $223S$ is very close to $242N$, but their *character* also repeats owing to the fact that $239A$ is also near to the magic number of days, 6,585 and a bit. Just what I mean by their character we will discuss later, but if we relax that added constraint, and look only for repeating occurrences of any sort, then we need to find only an agreement between S (the synodic month) and N (the nodical month).

Again a few strokes on the pocket calculator should satisfy you that

$$47 \times S = 1{,}387.94 \text{ days}$$
$$51 \times N = 1{,}387.82 \text{ days}$$

The difference amounts to about three hours.

This implies that an eclipse will likely take place 47 synodic months after a previous event. In terms of solar years that is a 3.8-year gap, almost exactly (I could have written 3.80005). Rather than convert the decimal to months and days it's easier just to count off the 1,388 days making up 3.8 years.

Again one can pore over tables of eclipses and check whether this is the case. I will not bore you with a whole string of examples, but take just one. Adding 3.8 years onto the July 6, 1982, total lunar eclipse invoked above, one expects a following eclipse about a week before the end of April in 1986. Sure enough, there was one on April 24.

There is an obvious relationship with the Metonic cycle here. Five multiplied by 3.8 equals 19 solar years, and 5 times 47 makes 235 synodic months. The 3.8-year cycle is a submultiple of the Metonic cycle. Not only do short sequences of eclipses occur with regular intervals of 19 years, but also that period is split up into five interleaving but distinct eclipse series.

The 3.8-year gap provides yet another regularity, then, which would allow investigators of eclipse records to make prognoses about future events once the pattern was recognized.

GEOGRAPHICAL SHIFTS IN ECLIPSE PATHS

So far we have concentrated on the spacing in time of eclipses. Next we consider some other characteristics. Flick back to Figure A-10. Imagine that saros A produced a total solar eclipse, so that the right-hand of the pair of disks may be thought of as equally well representing the Sun. Now think of the position of the Moon as it passed that position in saros B; that is, you slide it back down its inclined path until the two are aligned north–south, putting them at the same celestial longitude. In that position the center of the Moon is a little below that of the Sun, and so a total solar eclipse may still be witnessed in saros B, but its track on the Earth's surface will be displaced south from that which occurred 18.03 years earlier in saros A.

That is one distinct trend in eclipse occurrence representing a latitudinal shift. There will also tend to be an associated small shift in geographical longitude of the eclipse track because the terrestrial spin axis is tilted. However, there is another, larger, longitudinal shift, with a different origin. This was previously mentioned in Chapter 2.

The saros lasts for 6,585.32 days, indicating an excess of just less than one-third of a day over a round number of days. That represents almost a third of a rotation of the planet, the equivalent to 7 hours and 41 minutes. This means that the eclipse track is shifted by about 115 degrees to the west from one saros to the next.

These shifts—both north–south and east–west—were illustrated in Figure 2-2. Although the longitude movement is always from east to west, the latitudinal motion may be either from north to south or vice versa. In Figure 2-2, which is for total solar eclipses in saros sequence 136 during the twentieth century, the movement was from south to north because that sequence pertains to the descending node of the Moon's orbit. Other saronic sequences, associated with the ascending node (as in Figure A-10), demonstrate similar four-degree jumps, but from north to south.

A REPEAT ON REPEATING ECLIPSES

Referring again to Figure 2-2, the eclipses depicted all occurred around the middle of the year—shifting with forward steps of 10 or 11 days from May 18, 1901 to July 11, 1991, in accord with the saros—but are otherwise noteworthy because of the duration of totality. Most total solar eclipses last for only two or three minutes; the six eclipses shown each had totality lasting for about seven minutes. No natural solar eclipse will present such an opportunity again until the year 2150. One can increase the duration of totality by artificial means, by flying along the eclipse path as fast as you can in a supersonic aircraft, although even that cannot keep pace with the eclipse for much more than ten minutes.

When we first met the saros we merely noted that there was another near-coincidence with its length—that is, 239 anomalistic months last for 6,585.54 days, just 0.22 days longer than the saros—but we did not take that observation further at that stage.

The anomalistic month is the cycle time of the angular diameter of the Moon, altering between 0.548 degrees (at perigee) and 0.491 degrees (at apogee). At the end of a saros the Moon still has 0.22 days to go before it returns to the geocentric distance at which it began. If it started the saros precisely at perigee then at the end it has another 5 hours and 17 minutes to go before next passing through perigee. That is only one part in 125 of an orbit, the result being that the angular size of the Moon changes by very little, if measurements at start and completion of a saros are compared.

Now what about the apparent size of the Sun? That also affects whether or not an eclipse is going to total. The apparent solar diameter varies with the heliocentric distance of the Earth, and we saw earlier that it oscillates between 0.542 and 0.524 degrees during a complete orbit, or a full year. But we are not concerned with a full year. The saros lasts for 18.03 years implying that, compared with its beginning, at the end of a saros the Earth has traveled just 3 percent more than 18 complete orbits. Therefore the angular size of the Sun will not be much different from what it was at the start.

There is another remarkable coincidence, then. The apparent sizes of both Sun and Moon are close to being duplicated from one saros to the next. The eclipses in Figure 2-2 are a good example. Equally well the eclipses coupled with that of August 11, 1999, in saros 145 (those of July 31, 1981, and August 21, 2017, plus several others before and after) are also total eclipses, just shifted in steps west by 115 degrees and south by about 4 degrees. In the case of the 2017 eclipse this places the route beautifully across breadth of the contiguous United States, given that the 1999 event tracked over Europe and the Middle East.

Let us summarize what we have learned above. The saros enables us to predict repeating eclipses every 18.03 years, due to the fact that 223 synodic months happen to last for close to 242 nodical months. It also happens that 239 anomalistic months have essentially the same total duration, making the apparent size of the Moon not alter much after a saros, and the saros being not greatly different from 18 whole years results in the Sun also being near its original apparent diameter. These facts result not only in eclipses repeating, but also they repeat in basic *character*, a fact that was foreshadowed in Chapter 2 but not completely explained there.

HOW LONG IS THE PERIOD OF TOTALITY?

All the total eclipses in Figure 2-2 lasted for about seven minutes. What factors control that time span? The duration of totality depends upon the relative angular sizes of Sun and Moon. The greatest interval of obscuration is when a solar eclipse occurs (1) when the Moon is at perigee, so that the lunar diameter is maximized; and (2) when the Earth is at aphelion, so that the solar diameter is minimized (this is why those long eclipses straddled July, aphelion occurring early in that month).

The changing speeds of these bodies also affect the duration of totality: the apparent angular speed of the Sun is lowest when we are at aphelion, as above, and this enhances the duration of totality. On the other hand, when the Moon is at perigee its angular speed is the maximum it ever attains, and that has a contrary effect. Basically, seven-minute-plus eclipses result from the greatest feasible difference in lunar versus solar apparent diameter, about one-fortieth of a degree: Moon 0.548 degrees, Sun 0.524 degrees. The converse can also be true, the Moon appearing smaller than the Sun, making the duration of totality zero; that is, an annular eclipse occurs.

Apart from the stage of totality, we saw in Chapter 2 that there is an extended period—some hours—of partial eclipse that precedes and follows the main event. This is the time it takes for the Moon gradually to cover the Sun, and then to uncover it again later. More often, there is no totality (or even an annular eclipse) because the Moon does not pass centrally across the solar disk, and so only a partial eclipse takes place. Lunar eclipses may similarly be subdivided, and we should consider the different circumstances that can occur for those.

LUNAR ECLIPSE PHENOMENA

What basic phenomena occur during a lunar eclipse? A longitudinal section through the shadow cast by the Earth is shown in Figure A-12. If the Sun were a point object then the planet would produce only a complete shadow (termed the umbra), but the Sun is actually over half a degree wide. This makes the shadow fuzzy around the edges, producing a region called the penumbra.

This effect is easy to demonstrate in your back garden on any sunny day. Hold a sheet of paper up close to a shadow, such as that cast by the leaves of a tree. Near the leaves their shadows appear to have sharp, well-defined edges, but as you pull the paper back further they become less and less distinct. This is due to the finite size of the Sun.

Now consider the Earth in space rather than a leaf in your garden. The distance that the shadow is projected is immense. Figure A-12 is drawn in a much-compacted form: the angles between the straight lines are actually very small (about 0.533 degrees, that being the Sun's average angular diameter). This produces a conical shadow zone with the apex at point A, a distance 850,000 miles from the Earth.

If the Moon has a node close to opposition it will pass through

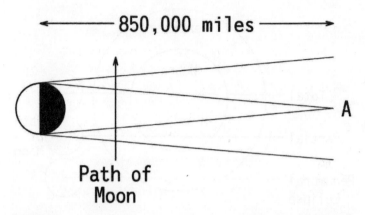

FIGURE A-12. The Earth casts a conical shadow that is about 850,000 miles long, to its apex labeled A here. When the Moon passes some- where through the shadow a lunar eclipse occurs (not to scale).

that shadow, and an eclipse will occur. The mean geocentric dis- tance of the Moon (238,850 miles) is about 28 percent of the distance to the apex of the shadow. As a result the umbra is 72 percent the diameter of the Earth at the position of the Moon, or about 5,700 miles across. Recall that the Moon is 2,160 miles in diameter, and so the umbra can easily envelop it. That is, a total lunar eclipse is easily achieved, and will last for some time as the Moon slowly moves through the umbra. On the other hand the penumbra is about 128 percent the planet's width at the lunar position, a diameter of close to 10,150 miles, and so almost five times the extent of the Moon. The sizes of the umbra and pen- umbra are portrayed to scale in Figure A-13 as slices through the terrestrial shadow, to show how total, partial, and penumbral lunar eclipses may occur.

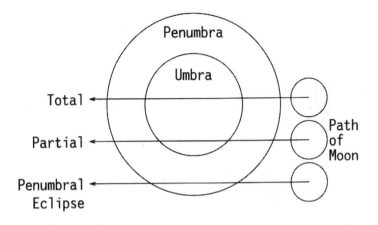

FIGURE A-13. A section through the terrestrial shadow shown in Figure A-12. The diameters of the Moon, umbra, and penumbra are shown to scale here. If the Moon completely enters the umbra, a *total* lunar eclipse occurs. A *partial* lunar eclipse is when only part of the lunar disk is enveloped in the umbra in any phase of the episode. The Moon passing wholly or in part through merely the penumbra is called a *penumbral* eclipse.

THE DURATIONS OF LUNAR ECLIPSES

How long do lunar eclipses last? How long does the Moon take to cross the umbra and the penumbra along paths like those shown in Figure A-13? The sums are quite easy to do once one knows the speed of the Moon in its orbit. (One might imagine that it is more complicated because the Earth's conical shadow is not staying still, moving along as the planet orbits the Sun, but remember that the Moon is moving with us.) A few taps on the pocket calculator show that the Moon's speed in its geocentric orbit is around 2,300 miles per hour, although variable between perigee and apogee. The diameter of the umbra is about 5,700 miles, so the Moon takes close to two and a half hours to traverse a central line through that shadow.

At least, that is what you get if you are considering just the center of the lunar disk. In reality, that is not what one observes. The Moon is large, and observers note when the edges of its apparent disk touch the extremes of the umbra and penumbra. As shown in Figure 2.5 it is conventional to define several distinct contact points. The first is P1, when the leading edge of the lunar limb touches the periphery of the penumbra. U1 is similarly defined for the initiation of entry to the umbra, and U2 is when the Moon is completely immersed therein. Exit from the umbra is U3, and then U4 is when the trailing part of the Moon escapes the umbra, the final exit from the penumbra being P4. (One could similarly define junctures P2 and P3 but they are of limited utility.)

The phase of totality for a lunar eclipse is between U2 and U3. This may last for an hour and a half, but it can be much less if the Moon crosses the umbra far off-center. Under such circumstances certainly most of the Moon is within the umbra for about an hour, but true totality is only briefly achieved. Unlike with total solar eclipses, the distinction is not important. The entire eclipse may be considered to last throughout the interval, with some part of the Moon within the penumbra, meaning from P1 to P4. This lasts for up to five and a half hours.

SOLAR ECLIPSE CONTACTS

Similar definitions to the above are used for defining the contact points during a solar eclipse, although the repeat usage of some of the alphanumeric terms can cause some confusion. P1 is when the partial solar eclipse begins, the lunar limb first appearing to touch the disk of the Sun, and similarly P2 is when the Moon wholly departs. For a total eclipse, U1 is defined as the instant at which totality begins, and U2 when it ends, the two being separated by merely a few minutes. For an annular eclipse, U1 is when the Moon is first completely enveloped within the solar disk, U2 when it touches the opposite solar limb.

THE ECLIPTIC LIMITS FOR LUNAR ECLIPSES

The fact that lunar eclipses are intrinsically less frequent than solar eclipses is reflected by the fact that the ecliptic limits are more stringent for the former. A total solar eclipse is *certain* if the Moon passes a node having a longitude within about 10 degrees of the Sun and *possible* if the separation is below about 12 degrees. For lunar eclipses one must compare the Moon's nodal longitude instead with the opposition point, 180 degrees away from the sunward direction. If the relevant gap is below 3.75 degrees then a total lunar eclipse is *certain*, and similarly *possible* beneath about 6 degrees. For partial lunar eclipses the corresponding ecliptic limits are 9.5 and 12.25 degrees.

The total lunar eclipse depicted in Figure 2-5 provides a good example. The Moon happens not to pass its node (that is, cross the ecliptic) until all phases of the eclipse are complete, that node being about 4 degrees from the opposition point (which is the middle of the umbra). That separation could have been considerably larger still, but again a total lunar eclipse would occur.

All the lunar ecliptic limits are substantially lower than the solar values, and that is why solar eclipses outnumber lunar eclipses by about three to two. Purely penumbral eclipses are more numerous, but often involve little more than a slight darkening of the full moon, and so we neglect them herein.

THE FREQUENCIES OF ECLIPSES

What is the maximum and minimum numbers of eclipses that can occur in any one calendar year? The matter of the minimum number is the easiest to address. The ecliptic limit pertaining to certainty of at least a partial solar eclipse is 15.35 degrees, producing a range in longitude of 30.7 degrees. The Sun appears to move along the ecliptic at just less than 1 degree per day (360 degrees to move and almost 365.26 days in a sidereal year). Therefore it takes just over 31 days to traverse the zone in which eclipses can occur, the seasons that recur twice per eclipse year when the lunar nodes are close to the solar direction. Because 31 days is longer than the nodical month, there must be at least one solar eclipse of some description in each eclipse season, making two each year. The eclipse year of 346.6 days most often is phased such that there are only two eclipse seasons in a calendar year, so that every calendar year must contain a minimum of two solar eclipses.

In contrast, partial lunar eclipses are certain only within ecliptic limits of 9.5 degrees, a range of 19 degrees in all, which the Sun takes just over 19 days to traverse, considerably less than a nodical month. Therefore it is possible for the Moon to avoid being eclipsed, in fact to avoid such ignominy in both eclipse seasons within a certain calendar year.

In consequence the minimum number of eclipses in any calendar year is *two*: both solar. Next we turn to the maximum.

The ecliptic limit rendering the possibility of partial solar eclipses is 18.5 degrees, making for a range of 37 degrees, which the Sun takes 37.5 days to traverse. One could get a solar eclipse at one conjunction, and then another at the following conjunction about 29.5 days later, both within that one eclipse season. Not only that, but a lunar eclipse between times is also feasible.

One can imagine, then, getting one lunar and two solar eclipses in an eclipse season, and in the next such season half an eclipse year later the same thing occurs, making six.

Is that the maximum? No, it's not quite. If the first eclipse season were centered on about January 15, the initial trio of eclipses would be in January with the lunar eclipse on that date and the initial solar eclipse on the first or second day of the month. The next set of three would be centered on July 8. Such a phasing allows for a third eclipse season partially lying within the calendar year, starting on December 12. A solar eclipse might occur soon thereafter, making seven in all within the calendar year, five solar and two lunar. In this scenario there cannot be a third lunar eclipse within the year, because twelve synodic months last for 354 days, and that period counted after January 15 puts any possible lunar eclipse twelve full moons later, on about January 4 of the *following* year.

A similar wrangling with dates allows one to ascertain that it is feasible to get four solar and three lunar eclipses in a year, again a total of seven. For this to occur one needs a lunar eclipse early in January followed by a solar eclipse at the next conjunction, then a solar/lunar/solar trio straddling the middle of the year, and finally in December a solar eclipse and paired lunar eclipse at the following opposition.

The bottom line is that in any calendar year there are at least two eclipses, both solar, but there may be up to a total of seven, split either 5:2 or 4:3 as solar:lunar. Nowadays that's of interest on a trivial level only, though, because such eclipses may be a mixture of partial, annular, and total, and for scientific purposes (and indeed public enthusiasm) it is really only the total eclipses that inspire. On the other hand, the mere keeping of records of when eclipses of any variety occurred would have allowed ancient civilizations to unravel the secrets of the cycles of the Moon. Our discussion of those cycles will have given you some inkling of how that could have been achieved.

THE DISTRIBUTION OF ECLIPSES

The average numbers of eclipses per century were mentioned in Chapter 2. The figures used were based on a monumental work by the nineteenth-century Viennese astronomer Theodor von Oppolzer, published posthumously in 1887. Using detailed theories for the orbits of the Sun and Moon, Oppolzer calculated by hand the circumstances for all eclipses between 1208 B.C. and A.D. 2161, a total of 3,368 years providing in all 8,000 solar and 5,200 lunar eclipses. From this compendium are derived the averages of 238 solar and 154 lunar eclipses per century.

These may further be subdivided into partial and total events, and so on. Easiest to analyze are the lunar eclipses: over a hundred years about 71 total and 83 partial lunar eclipses may be expected.

Turning to solar eclipses, the 238 per century break down as 84 partial, 66 total, 77 annular, and 11 partly annular and partly total.

How could a particular eclipse event be both? Consider Figure 2-1 again. The nearest part of the Earth's surface to the Moon, around the noon meridian, may be only just close enough to be within the umbra (the conical lunar shadow), so that observers there experience a very brief total eclipse. Further to the east and the west the observers are a few thousand miles more distant, putting them beyond the vertex of the umbral cone, so that they witness only an annular eclipse. The track of the eclipse drawn across the globe would start in the west as an annular phenomenon, become total as the point of greatest eclipse is approached, and then become annular again as the track proceeds east. This may be termed a *hybrid eclipse*.

If there are 66 total eclipses per century, then such an opportunity presents itself somewhere around the world once every 18 months on average. If you were clever enough to take advantage of one of the 11 hybrid eclipses by placing yourself within the portion of the ground track achieving totality, then with an unlimited travel budget you might manage one total eclipse every 15 or 16 months, on average. They are not smoothly distributed in time, though.

Unfortunately many total solar eclipses have paths unfavorable for potential viewers, and a track traversing an accessible location with a good chance of clear weather occurs only about once every three years. Nevertheless, many is the keen eclipse watcher who has spent an enormous amount of time and money getting to a well-considered prime spot, only to be stymied by an unseasonably cloudy day.

These numbers of eclipses per century are all averages, such as would result if they happened randomly in time. But we know that is not reality. They repeat on regular cycles. Total solar eclipse tracks perform consistent geographical steps within a saros, as in Figure 2.2, and there are systematic trends in other eclipse sequences.

There is another geographical effect that we have yet to mention, although it was alluded to at the start of this book. Taking into account the summed area of a track of totality across the surface of the Earth, and the average occurrence rate, for any random point on the planet a total solar eclipse might be expected about once per 410 years. But just as they are not randomly distributed in time, so they do not occur randomly in terms of geography.

A total solar eclipse is more likely to happen while the Earth is near aphelion than when near perihelion, because while we are further from the Sun its apparent diameter is minimized, presenting less of a target area for the Moon to obscure. This means that more total solar eclipses occur between May and August (straddling aphelion in early July) than between November and February (bracketing perihelion in early January), at least in the present epoch. Over the next six or seven millennia the date of perihelion will move much later in the year, eventually reversing this trend.

This implies that more total eclipses occur during the Northern Hemisphere summer than its winter. Summer is the time when the Northern Hemisphere is tipped over towards the Sun (that's why it *is* summer), as in Figure A-3, presenting a larger sunward area than the Southern Hemisphere. Overall the effect is that the north gets more total solar eclipses. Averaged over the globe the rate is about one per 410 years for a random location, but a random location chosen in the Northern Hemisphere gets one total eclipse every 330 years or so, whereas in the Southern Hemisphere it is less frequent, once per 540 years.

As the bulk of the population lives in the Northern Hemisphere, a person picked at random from the whole of humankind has an enhanced probability of experiencing a total solar eclipse without needing to chase after one. Lifetimes average to about 80 years in the developed world, such as in North America, Europe, or Japan. A randomly chosen person from such a country therefore has about a one-in-four chance of happening to be crossed by a total solar eclipse track during his or her lifetime.

That probability can be turned into a certainty by going in chase of such an event. I hope this book will have persuaded you that this is an attractive idea.

Glossary of Astronomical and Scientific Terms

Albedo The fraction of impinging sunlight reflected by a celestial body.

Aphelion The greatest distance in its orbit of the Earth (or any other celestial body) from the Sun.

Apogee The greatest distance in its orbit of the Moon from the Earth; may also be applied to other objects, such as artificial satellites.

Appulse When two celestial bodies come into conjunction but do not quite eclipse, occult, or transit each other (e.g., an asteroid passing very close by a star in the sky).

Arcsecond A measure of angle equivalent to one-sixtieth of an arcminute, which in turn is one-sixtieth of a degree, there being 360 degrees in a complete circle.

Astronomical Unit (AU) The average distance between the Earth and the Sun (1 AU is approximately 93 million miles, or 150 million kilometers).

Azimuth The angular distance measured along the horizon from a fixed point (usually clockwise from due north).

Baily's beads The visual phenomena seen just before or during totality in a solar eclipse, the light from the photosphere reaching the eye through valleys around the periphery of the Moon appearing to form moving beads.

Barycenter The combined center of mass of two or more orbiting bodies.

Celestial equator The equatorial plane of the Earth extrapolated out into the sky.

Celestial latitude The angle north or south of the ecliptic plane.

Celestial longitude The angle around the ecliptic from the spring equinox position to some specified point; this is measured counterclockwise (i.e., in the direction of the orbital motion of the Earth, the Moon, and the other planets).

Chord The path apparently taken by an eclipsing body across an eclipsed body (e.g., Venus or Mercury in transit across the Sun).

Chromosphere A layer a few thousand miles thick between the photosphere and the corona that may be seen fleetingly during an eclipse as a circle of red spikes around the outside of the Sun.

Coma The vast cloud of gas and dust around the solid nucleus of a comet formed when sunlight causes some of its icy content to evaporate.

Conjunction The alignment of two celestial bodies when they are at the same celestial longitude. Lunar **superior conjunction** (i.e., when the longitudes are 180 degrees different making the Moon full, such that a lunar eclipse may occur) is usually termed **opposition**. **Inferior conjunction** (when the Moon may eclipse the Sun) is often simply termed **conjunction**; many accounts incorrectly refer to this as being the time of new moon.

Contact points The points (and instants of time) when the silhouettes and shadows of eclipsing bodies come into contact; these define the periods of the total and partial phases of an eclipse. The contact points may be either internal or external (e.g., when Venus touches the solar disk on the outside, and when it wholly enters that disk a short time later) leading to the concepts of ingress and egress, or immersion and emersion.

Corona The outermost layer of the solar atmosphere, spreading millions of miles out into space, which may be seen as a white halo around the Moon during a total solar eclipse.

Diamond–ring effect The visual phenomenon usually seen just before totality is reached in a solar eclipse, the corona and chromosphere providing the apparent ring while the final visible part of the photosphere sparkles like a diamond set on that ring.

Eclipse season A period of time during which the Sun (or the Moon) is between the ecliptic limits.

Eclipse year The time between successive passages of the Sun through the lunar nodes; duration 346.6 days. Eclipses of the Sun and the Moon can only occur during the eclipse seasons that straddle the middle and the beginning/end of each eclipse year.

Ecliptic The plane of the Earth's orbit, and also the apparent path of the Sun across the sky.

Ecliptic limits The range of possible celestial longitudes within which an eclipse can occur.

Flare A massive ejection of solar material into space, seen as a bright globule. Larger flares appear as prominences during an eclipse.

Ground track The shadow of an eclipsing body drawn across the surface of the Earth.

Inclination The tilt of the Moon's orbit relative to the ecliptic. (May also be applied to other objects such as planets, comets, and asteroids.)

Ion An atom or molecule with either fewer or more electrons than usual, giving it a net positive or negative electrical charge.

Light–curve A graph of the way in which the brightness of a body changes as it is eclipsed, or otherwise changes in intensity (e.g., variable stars display intrinsic cyclic changes in brightness).

Metonic cycle A period of 19 solar years, which is very close to 235 synodic months in duration. Apart from providing for eclipse sequences, the Metonic cycle is used in various calendar schemes (e.g., the calculation of dates of Easter).

Node Either of two diametrically opposite points where the orbit of the Moon, a satellite, a planet, or some other celestial object crosses the ecliptic. If passing from south of the ecliptic to the north that point is termed the ascending node; the other is the descending node.

Occultation The eclipse of a star or some other distant body such as a galaxy or quasar by the Moon or another solar system body (a planet, asteroid, or comet).

Opposition The time at which a celestial object is opposite the Sun in the sky such that their celestial longitudes differ by 180 degrees; see also **Conjunction**.

Penumbra The area of partial shadow surrounding that of complete shadow (the **umbra**) in an eclipse.

Perigee The closest approach in its orbit of the Moon to the Earth; may also be applied to other objects such as artificial satellites.

Perihelion The closest approach in its orbit of the Earth (or any other celestial body) to the Sun. See also **Aphelion**.

Photosphere The visible surface of the Sun. During a total eclipse the photosphere is obscured by the Moon, allowing the chromosphere, corona, and prominences to be seen.

Precession The gradual shift in some celestial angular measure due to gravitational perturbations. For example, the Earth's spin axis precesses in a clockwise direction under the influence mainly of the Moon and the Sun, taking 25,800 years to complete a 360 degree turn, and this results in the precession of the equinoxes. The perihelion point of the Earth's orbit around the Sun precesses in the counterclockwise direction under gravitational tugs by the other planets, taking 110,000 years to make a full revolution.

Prominence A cloud of gas protruding outwards from the chromosphere into the corona. Prominences are spectacular reddish structures often seen during total solar eclipses; they may be over 100,000 miles high.

Redshift The displacement of spectral lines toward the red end of the spectrum, due to an object's recessional speed; caused by the Doppler effect.

Refraction The bending of light rays as they pass from one medium (e.g., air) into another of different density (e.g., glass).

Saros The cycle of 18 years plus 10 or 11 days over which eclipses repeat. The saros is due to the near synchronicity of integer multiples of the synodic, anomalistic, and nodical months, as explained in the Appendix.

Seeing A measure of the twinkling of stars caused by atmospheric turbulence. At any particular observatory the seeing may vary from hour to hour and night to night, and is measured in **arcseconds**.

Shadow bands Short-lived alternating bright and dark bands of light that may be seen on the ground and other suitable structures in the last few seconds before a solar eclipse becomes total. Their origin is similar to the **seeing** of stars, being caused by the propagation through the turbulent atmosphere of light from a small source, in this case the very slender crescent of the solar disk shortly before it is obscured.

Spicules The numerous short, spiky prominences seen during an eclipse to project up from the chromosphere into the lower corona.

Sunspot A dark patch seen on the photosphere, slightly cooler than its surrounds.

Synodic month The time from one full moon to the next; also known as a **lunation**. There are other types of lunar month (sidereal month, anomalistic month, draconic or nodical month), as discussed in the Appendix.

Syzygy Either of the two points in the lunar orbit that are aligned with the Earth and Sun in terms of celestial longitude (i.e., inferior and superior conjunction, or opposition), where an eclipse is possible.

Transit The passage of Mercury, Venus, or some other celestial body (e.g., an asteroid) across the face of the Sun. May also be applied to other cosmic situations (e.g., transit of a moon across the face of Jupiter, or transit by one star in a binary pair across the other).

Umbra The region of complete shadow in an eclipse.

Universal Time (UT) The reference system used to coordinate most time-keeping worldwide, especially in astronomical observations. It may be thought of as being the mean solar time for the Greenwich meridian (i.e., GMT) although the definition is slightly different.

Zenith The point in the sky directly above the observer.

Picture Credits

The following sources are gratefully acknowledged. All other diagrams and photographs are by the author.

Figure 1-1: National Solar Observatory/Sacramento Peak, Sunspot, New Mexico.

Figure 1-2: John Walker, Fourmilab, Switzerland.

Figure 1-3: High Altitude Observatory, National Center for Atmospheric Research, Boulder, Colorado.

Figures 1-4, 13-3, and 13-4: G.F. Chambers, *Handbook of Descriptive Astronomy* (Clarendon Press, Oxford, 1877).

Figure 1-5: SOHO/Extreme Ultraviolet Imaging Telescope consortium. SOHO is a project of international cooperation between ESA and NASA.

Figure 1-6: NASA and the crew of Apollo 12.

Figure 1-7: Royal Astronomical Society, from the *Memoirs*, 1836.

Figure 1-8: NASA and the crew of Apollo 11.

Figure 1-9: John Kennewell, Learmonth Solar Observatory, Western Australia.

Figures 1-10, 1-11, and 3-5: J.F. Blake, *Astronomical Myths, Based on Flammarion's "History of the Heavens"* (MacMillan, London, 1877).

Figures 1-12, 1-14, 3-3, 3-4, 7-1, 7-2, 9-2, and 15-4: Royal Astronomical Society.

Figure 1-13: Simon Newcomb, *Popular Astronomy* (MacMillan, London, 1878).

Figures 2-2, 4-3, 4-4, and 5-7: F. Dyson and R.v.d.R. Wooley, *Eclipses of the Sun and Moon* (Clarendon Press, Oxford, 1937).

Figure 2-4: CNES (France) and the crew of Mir mission 27.

Figures 2-5, 15-7, and 15-8: Fred Espenak, NASA-Goddard Space Flight Center, Greenbelt, Maryland.

Figure 2-6: M. Payen, *Selenelion ou Apparition Luni-Solaire en l'Isle de Gorgone*, 1666. Reproduced in *L'Astronomie* (Paris), volume 12, pp.13-18, 1893; courtesy W.G. Waddington, University of Oxford.

Figure 2-7: John Hisco, Astronomical Society of South Australia.

Figure 3-6: H. Rider Haggard, *King Solomon's Mines* (Cassell, London, 1886).

Figure 4-2, H. Spencer Jones, *General Astronomy* (Edward Arnold, London, 1922).

Figures 4-5 and 4-6: S.A. Mitchell, *Eclipses of the Sun* (Columbia University Press, New York, 1923).

Figure 4-8: Geraint Lewis and Michael Irwin; William Herschel Telescope, La Palma.

Figure 4-9: Warrick Couch, Richard Ellis, Space Telescope Science Institute, and NASA.

Figure 5-2: Courtesy of the SOHO/LASCO consortium. SOHO is a project of international cooperation between ESA and NASA.

Figure 5-3: Clementine image courtesy U.S. Naval Research Laboratories.

Figure 5-4, 5-5, and 5-6: Norman Lockyer, *Recent and Coming Eclipses* (MacMillan, London, 1900).

Figure 6-1: NASA and the crew of Apollo 14.

Figure 6-2: Courtesy Steven Bell (H.M. Nautical Almanac Office), and the Royal Astronomical Society.

Figure 8-1: Washington Irving, *Life and Voyages of Christopher Columbus* (Philadelphia, 1892).

Figure 8-2: Thomas McKenney and James Hall, *History of the Indian Tribes of North America* (Philadelphia, 1837-1844).

Figure 9-1: J.J. Grandville, *Un Autre Monde* (Paris, 1844).

Figure 10-1: *Cortland* (New York) *Evening Standard*, May 28, 1900.

Figure 10-2: *The Bell System Technical Journal*, volume 27, pp. 510-588, 1948.

Figure 10-3: Courtesy Yale University, New Haven, Connecticut.

Figure 11-1: Courtesy the Maria Mitchell Association, Nantucket, Massachusetts.

Figure 12-1: Applied Physics Laboratory, Johns Hopkins University, and NASA.

Index

Hydrogen
  coma and, 260
  masers, 147
  in stars, 125–126
  in Sun, 5, 127, 130, 142, *143,*
    146, 160, 281

I

Ice Age, 154–155, 157–158, 394
Ida (asteroid), 320
Immersion, 282
Incarnation, 367
Inclination, 43, 402, *404, 413,* 437
India, Indians, 16, 17, 86, 136
Indians, American, 175, 188. *See also*
    Native Americans; *specific*
    Indian tribes
Industrial Revolution, 262
Ingress, 282, *283,* 292, 295, 296
International Astronomical Union,
    234
Internet, xiv
Io, *303*
Ionia, Ionians, 382
Ionosphere, 133
Ion(s), 144–145, 260, 437
Iran, Iranians, 393, 394
Ireland, Irish, 20, 361, 366, 372, 375,
    378
Irving, Washington, 98, *177*
Isandlwana, Battle of, 210–212
Ishtar, 28

J

Jacquard, Joseph-Marie, 73
Jamaica, Jamaicans, 96–98, *99, 177,*
    188
Janssen, Jules, 135, 143, 144
Jefferson, Thomas, 191–192
Jeffrey, Lord Francis, 245
Jesus Christ, vii, 1, 20–23, 367
Jews, Judaism, xiii, 21, 22, 28, 85,
    365
Joshua, 19–20
Joslin, Rebecca R., 31–32
Joyce, James, 185
Judea, Judeans, 20, 85
Jupiter
  asteroids near, 254
  comets and, 260
  energy of, 108
  Galileo and, 320
  Mars and, 21, 303–304
  Mercury and, 305
  moons of, 92, 207, 289, 297, 298,
    302, *303,* 314, 384
  Pluto Express and, 316
  ring around, 270
  Saturn and, 21
  seven-day week and, 85
  size of, 160
  Sun and, 126
  during total solar eclipse, 335
  Uranus and, 265
  Venus and, 302, 305
Justinian, 362
Jutes, 360, 361, 363

Morrison, Leslie, 155, 172
Motion, laws of
    Kepler's, 314
    Newton's, ix, 178
Mount Pinatubo, 63
Mount St. Helens, 63
Müller, Johannes. *See*
        Regiomontanus

N

Nanometer, 145
NASA (National Aeronautics and
        Space Administration)
    Ames Research Center, 257
    Goddard Space Flight Center,
        xiv
    Langley Research Center, 201
    Mariner 10 satellite, 315
    NEAR-Shoemaker satellite, *255*
    Solar and Heliospheric
        Observatory, 133
    Voyager 2, 266, 267
Native Americans, 175. *See also*
        Indians, American; *specific
        Native American tribes*
Nativity, 21
Nature, 202
Nautical Almanac Office, U.S., 233
Naval Observatory, U.S., 200, 202,
        217
Navigation, 289
    Bowditch on, 187
    Moon and, 91–93, 285
    Regiomontanus and, 94, 96

solar wind and, 131
transits of 1761 and 1769 and,
        299
Venus and, 301
Navy, Royal, 74, 294
Navy, U.S., *118,* 225–226
NEAR-Shoemaker satellite, *255*
Nebuchadnezzar, 84–85
Nebulae, 126, 263
Nebulium, 144–145
Neptune
    Adams and, 269
    energy of, 108
    Le Verrier and, 269, 304, 305–
        306
    Mercury and, 305
    minor planets and, 317
    moon of, 384
    orbit of, 312
    Planet X and, 317
    Pluto and, 309, 310
    rings around, 267, 269–270, *271*
    Uranus and, 311–312
New American Practical Navigator, The
        (Bowditch), 184
Newcomb, Simon, 295, 299
New English Bible, vii, 22
New moon
    conjunction and, 42
    Easter after, 366, 370, 375, 376,
        379, 380
    fertility near, 38
    Islamic calendar and, 40
    Metonic cycle and, 44
    Sun and, *401*
    during Zulu Wars, 211

Precession
definition of, 44–45, 438–439
Earth's orbit and, 400, 401
of equinoxes, 81, 82, 403
of Mercury, 306, 308
of nodes, 403
of perigee, 403
of perihelion, 394–397
of Sun, 276
of Venus, 276
Princeton University, 200
Printing, 94
Procyon, 209, 335
Prominence(s), 8–10, 51, *128*
of 664, 373
definition of, 130, 439
during 1878 eclipse, 209
ground track and, 182
monitoring of, 133
in nineteenth century, 138
rain and, 349
spectrum of, *143*
during totality, 338
Protestant Church, 368
Protons, 127
Ptolemies, 29, 78, 86
Pydna, Battle of, 17, 35
Pythagoras, 29

Q

Quantum theory, 145
Quasar(s), *122,* 123–124, 151, 251–252, 272
Quasi-stellar object. *See* quasar(s)

R

Radiative zone, 128
Radio communications, 6, 131, 252–253
Rain, 349
Raleigh, Sir Walter, 95
Records, recordkeeping, 147–149
by Babylonians, 14–16, 85, 86–87
in Bible, 19–20
by Chinese, 16, 155–156
of comet, 72
eclipse cycles from, 66, 74–78, 431
by Greeks, 16
in monasteries, 361
year length from, 28
Redshift, 251, 439
Reformation, 368
Refraction, 260, 439. *See also* telescope(s)
Regiomontanus, 94, 96
Regulus, 209, 245, 346
Relativity, general theory of, x, 70, 104, 109–113, 119, 308
Rémi, Georges, 103
Retro-reflector, laser, 149, 157
Revolutionary War, 178
Rheticus, Georg Joachim, 273
Rider, Barbara, 222
Rio Grande railroad, 200
Rocky Mountain News, 202, 208, 209
Roman Catholic Church, 368, 390, 393
Roman Church, 358, 360–379, 407